The Essential Guide to Creating Multiplayer Games with Godot 4.0

Harness the power of Godot Engine's GDScript network API to connect players in multiplayer games

Henrique Campos

BIRMINGHAM—MUMBAI

The Essential Guide to Creating Multiplayer Games with Godot 4.0

Group Product Manager: Rohit Rajkumar

Publishing Product Manager: Vaideeshwari Muralikrishnan

Book Project Manager: Sonam Pandey

Content Development Editor: Debolina Acharyya

Technical Editor: Reenish Kulshrestha

Copy Editor: Safis Editing

Proofreader: Safis Editing

Indexer: Tejal Daruwale Soni

Production Designer: Prashant Ghare and Jyoti Kadam

DevRel Marketing Coordinators: Nivedita Pandey, Namita Velgekar, and Anamika Singh

First published: December 2023

Production reference: 2201223

Published by Packt Publishing Ltd.
Grosvenor House
11 St Paul's Square
Birmingham
B3 1RB, UK

ISBN 978-1-80323-261-4

www.packtpub.com

To God, who has always been me with me, blessing me with strength, faith, and determination to take faithful steps throughout this journey. Let this book be a blessing that will allow people to make amazing creative endeavors.

– Henrique Campos

Foreword

Henrique, better known as Pigdev in the Godot sphere, was one of the first content creators covering the engine back in the day. He's always been dedicated to open source software.

We worked together for a couple of years, and I'm glad to see how Henrique has grown into one of the few old-timers who are still constantly learning and sharing with the community.

Most resources for Godot focus on helping newcomers learn the basics. With this essential guide to creating multiplayer games with Godot, we finally have a learning resource dedicated to online multiplayers. At the time of release, it is one of the areas of Godot that lacks learning resources the most.

I especially appreciated the last three chapters, as they dive into essential techniques such as lag compensation and optimizing network requests. Serious real-time multiplayer games absolutely need them, yet information on these topics can be especially hard to find.

I hope you'll have a great time with this essential guide to multiplayer in Godot, and I can't wait to see more multiplayer Godot games thanks to it.

Nathan Lovato

GDQuest founder

Contributors

About the author

Henrique Campos is an indie game developer and game designer working in the industry since 2015. Starting as a university teacher in the Computer Graphics and Artificial Intelligence chairs and working for the GDQuest team from 2018 to 2022, he also provides consultancy for solo developers, studios, and schools.

Under the alias of Pigdev, Henrique has been creating game development content on his YouTube channel since 2016. Among his projects, he wrote the *Top 7 Godot Engine Recipes* and *Platformer Essential Recipes* e-books, where he presents design patterns that people can use to make games with the Godot Engine.

A passionate open source enthusiast, Henrique has been working and contributing to the Godot Engine project since 2016.

I want to thank God, my friends, and my family, especially my sister, Ráisa, my mother, Sheila, and my father, Claudio, for always being by my side and supporting me throughout my journey of empowering indie game developers.

I also want to thank the whole Godot Engine community for building and supporting such an amazing tool.

About the reviewer

Yogendra Manawat is an indie maker and computer science student. While he is primarily focused on development skills, he's also experienced in programming languages including C++, Python, C#, TypeScript, and GDScript.

His passion for games and movies led him to explore 3D animation with Blender. Yogendra has over three years of game development experience, having created several small games. Currently, he's focused on his original story-based game, Bleak, built using the Godot Engine.

Yogendra loves learning new technologies and experimenting with development techniques to create unique games and useful software. He's an active contributor to the Godot Engine community and has developed tools, including Coursefy and Vortexorama, using Godot. Yogendra actively engages in game development forums, collaborating and exchanging ideas with fellow developers.

Table of Contents

3

Making a Lobby to Gather Players Together   47

4

Creating an Online Chat   73

Part 2: Creating Online Multiplayer Mechanics

5

Making an Online Quiz Game   83

6

Building an Online Checkers Game   101

7

Developing an Online Pong Game   127

8

Creating an Online Co-Op Platformer Prototype   145

Preface

The Essential Guide to Creating Multiplayer Games with Godot 4.0 is the ultimate hands-on guide to understanding how to make online multiplayer games using the open source Godot Engine. In its fourth version, Godot Engine introduces a high-level network API that allows users to focus on creating interesting and funny mechanics while letting the engine do the heavy work.

Through this book, you will learn the fundamentals of networks, including the basic UDP, TCP, and HTTP protocols. You will see how Godot Engine uses its ENet library implementation to seamlessly integrate these protocols into its game development workflow. Through nine projects, including five games, one of them an online multiplayer adventure game, you will learn all you need to know to connect players together in amazing shared experiences.

Who this book is for

This book is made for intermediary Godot Engine users, people who already know how Godot Engine works, its design philosophy, the editor, the documentation, and its core features. These users have already made games using Godot Engine. They are looking for something to make their next project stand out, and adding online multiplayer features is that thing.

What this book covers

Chapter 1, Setting up a Server, explains what a network is and how Godot Engine implements networking features through its ENet library implementation. You'll learn how to make your first handshake to effectively connect a server and a client machine.

Chapter 2, Sending and Receiving Data, discusses the foundation of a network, which is for multiple computers to exchange data with each other. In this chapter, you will learn that we use a data structure called packets, and we serialize data to recreate game states on the network. For this, we use the UDP protocol. At the end, we will have a login screen, effectively retrieving data from the server.

Chapter 3, Making a Lobby to Gather Players Together, explains how Godot Engine eases the process of serializing and exchanging data using the industry-standard UDP protocol by providing **Remote Procedure Calls** (**RPCs**), allowing us to essentially make calls to methods on remote objects. By the end of the chapter, we expand the login screen into a lobby, adding another client to the equation and connecting three computers together.

Chapter 4, Creating an Online Chat, explains that with the power of RPCs, we can now change objects' states remotely with ease. In this chapter, we learn how we use RPCs to allow players to exchange messages in a chat. We discuss how we can use RPCs and channels to prevent network bottlenecks. With that, we prepare ourselves to implement actual game features. By the end of the chapter, we will have a fully functional chat.

Chapter 5, Making an Online Quiz Game, explains how we can synchronize players' game states based on the general game state. We will set up a server that will react to player interactions and change the game state from waiting for a response to processing a match winner, announcing the winner, starting a new match, and reaching the end of available quiz questions, effectively ending the game. By the end of the chapter, we will have an online quiz game where multiple players compete to answer the most questions correctly.

Chapter 6, Building an Online Checkers Game, moves on to implementing a turn-based multiplayer online game, and nothing is better than the classic checkers for that. In this chapter, we will learn how to get the most out of our RPCs while still maintaining the heavy processing on players' machines. We will also discuss MultiplayerSynchronizer, a node that allows us to easily synchronize nodes' properties remotely. We will also learn what Multiplayer Authority is, which prevents players from messing around with other players' objects. By the end of the chapter, we will have a fully functional online checkers game.

Chapter 7, Developing an Online Pong Game, begins the transition from turn-based to action. Action games rely heavily on players' reaction times, and the game world should update its state quickly to allow players to have a smooth experience. Here, we will develop an online Pong game and use the MultiplayerSynchronizer node to sync the players' paddles and the ball. We will also learn that some features should use different syncing processes. We will go even deeper into the Multiplayer Authority realm to prevent one player's input from interfering with the other player's paddle's movement. By the end of the chapter, we have a playable multiplayer online Pong game.

Chapter 8, Designing an Online Co-Op Platformer, is where our baby steps stop and we start to implement interesting features for a custom game. We will prototype a physics puzzle platformer game where players grab and move crates around to overcome obstacles and reach the level's goal. Applying everything we've learned in the previous chapters, we will expand the usage of the MultiplayerSynchronizer by syncing animations on top of objects' positions. By the end of the chapter, we will have a working prototype of a physics puzzle co-op platformer game.

Chapter 9, Creating an Online Adventure Prototype, is, if we're honest, what you have been looking for throughout this book. Here, we will use all our skills to create an online multiplayer adventure game. Players can join and leave at any time and the world is persistent, maintaining players' progress in quests. We will discuss the basics of making an MMORPG game, storing and retrieving players' quest progress in a database, syncing players' spaceships and bullets, and making their actions affect other players' experiences. By the end of the chapter, we will have a prototype of an online multiplayer top-down space-shooter adventure game.

Chapter 10, Debugging and Profiling the Network, moves on from implementing the online multiplayer features on our top-down space-shooting adventure prototype. We now need to pave the way for thousands of players to play our game simultaneously. For that, we will use Godot Engine's built-in debugging and profiling tools to assess the potential areas for improvements in our game. We will focus on the network profiler and the monitor debugging tools to assess and propose potential solutions for the bottlenecks we find in our prototype. By the end of the chapter, we will have two of the most powerful and necessary skills a developer can have: the ability to debug and optimize a game.

Chapter 11, Optimizing Data Requests, builds on the understanding of the tools we have at our disposal to assess the information we need to discover potential areas for improvement; now, it's time to get our hands dirty. Throughout this chapter, we will learn how to create custom monitors to gather data about specific game features and decide the best strategy to optimize them. By the end of the chapter, we will have refactored our top-down space-shooting adventure, decreasing the bandwidth and the number of RPCs we make, effectively making our network consumption much lighter. We will have also implemented several techniques to decrease the network load, assessing each improvement with the network profiler and custom monitors to see how much better the game is becoming.

Chapter 12, Implementing Lag Compensation, deals with the issue that due to the improvements made to decrease network usage, our game may be inaccurately replicated on players' machines. With fewer RPCs and more sparse synchronization, the game may become asynchronous among players. Add latency and packet loss on top of that and you effectively worsen the players' experience. Nobody likes lag in their game. In this chapter, we will learn how to use Tweens to implement interpolation, prediction, and extrapolation to compensate for all these issues. By the end of the chapter, we will have a version of the top-down space-shooting adventure prototype with some fake latency and solutions for this game-breaking issue.

Chapter 13, Caching Data to Decrease Bandwidth, handles an important issue: throughout our network engineering endeavors, we have learned that bandwidth is our core resource and that we should always look to optimize its usage. In this chapter, we will learn how to use HTTP to download some data and store it on players' machines so that we can reuse it when necessary. By the end of the chapter, we will have implemented a feature that allows players to use custom images for their spaceships, and this new image will be replicated on all other players' instances of the game. To save bandwidth, we will implement caching using the user data folder.

To get the most out of this book

This book uses Godot Engine's ENet library implementation to create several prototypes that explore the boundaries of this technology. To get the most out of this book, you need to understand how Godot Engine works, how to code in GDScript, how to use Git, and the basics of the UDP, TCP, and HTTP protocols.

Software/hardware covered in the book	Operating system requirements
Godot Engine 4.0	Windows, macOS, or Linux

Throughout the book, we use the projects available in the book's GitHub repository. This book focuses on the networking aspects of each project, so to save you time, you have ready-to-use projects available so you don't need to bother with implementing other features that are not related to network engineering. In that sense, having Git installed is recommended, although not mandatory because you can download the code directly through the links provided as well.

If you are using the digital version of this book, we advise you to type the code yourself or access the code from the book's GitHub repository (a link is available in the next section). Doing so will help you avoid any potential errors related to the copying and pasting of code.

Download the example code files

You can download the example code files for this book from GitHub at `https://github.com/PacktPublishing/The-Essential-Guide-to-Creating-Multiplayer-Games-with-Godot-4.0/`. If there's an update to the code, it will be updated in the GitHub repository.

We also have other code bundles from our rich catalog of books and videos available at `https://github.com/PacktPublishing/`. Check them out!

Conventions used

There are a number of text conventions used throughout this book.

`Code in text`: Indicates code words in text, database table names, folder names, filenames, file extensions, pathnames, dummy URLs, user input, and Twitter handles. Here is an example: "One of the Godot Engine's Network API core features is the `ENetMultiplayerPeer` class."

A block of code is set as follows:

```
func _process(delta):
    server.poll()
    if server.is_connection_available():
        var peer = server.take_connection()
        var message = JSON.parse_string(peer.get_var())
        if "authenticate_credentials" in message:
            authenticate_player(peer, message)
        elif "get_authentication_token" in message:
            get_authentication_token(peer, message)
```

Bold: Indicates a new term, an important word, or words that you see onscreen. For instance, words in menus or dialog boxes appear in **bold**. Here is an example: "From there, we first need a server. So, choose one instance and press the **ServerButton**."

> **Tips or important notes**
> Appear like this.

Get in touch

Feedback from our readers is always welcome.

General feedback: If you have questions about any aspect of this book, email us at customercare@ packtpub.com and mention the book title in the subject of your message.

Errata: Although we have taken every care to ensure the accuracy of our content, mistakes do happen. If you have found a mistake in this book, we would be grateful if you would report this to us. Please visit www.packtpub.com/support/errata and fill in the form.

Piracy: If you come across any illegal copies of our works in any form on the internet, we would be grateful if you would provide us with the location address or website name. Please contact us at copyright@packt.com with a link to the material.

If you are interested in becoming an author: If there is a topic that you have expertise in and you are interested in either writing or contributing to a book, please visit authors.packtpub.com.

Share Your Thoughts

Once you've read, we'd love to hear your thoughts! Scan the QR code below to go straight to the Amazon review page for this book and share your feedback.

https://packt.link/r/1803232617

Your review is important to us and the tech community and will help us make sure we're delivering excellent quality content.

Download a free PDF copy of this book

Thanks for purchasing this book!

Do you like to read on the go but are unable to carry your print books everywhere?

Is your eBook purchase not compatible with the device of your choice?

Don't worry, now with every Packt book you get a DRM-free PDF version of that book at no cost.

Read anywhere, any place, on any device. Search, copy, and paste code from your favorite technical books directly into your application.

The perks don't stop there, you can get exclusive access to discounts, newsletters, and great free content in your inbox daily

Follow these simple steps to get the benefits:

1. Scan the QR code or visit the link below

https://packt.link/free-ebook/9781803232614

2. Submit your proof of purchase
3. That's it! We'll send your free PDF and other benefits to your email directly

Part 1: Handshaking and Networking

Throughout this part of the book, we take our first steps into the realm of networking. We start by making a handshake using Godot Engine's high-level `EnetMultiplayerPeer` class. We also learn how to use the UDP protocol to exchange data and end up by learning how to use **Remote Procedure Call (RPC)**.

This part contains the following chapters:

- *Chapter 1, Setting up a Server*
- *Chapter 2, Sending and Receiving Data*
- *Chapter 3, Making a Lobby to Gather Players Together*
- *Chapter 4, Creating an Online Chat*

1

Setting up a Server

Welcome to *The Essential Guide to Creating Multiplayer Games with Godot 4.0*. In this hands-on book, you are going to learn the core concepts used to create online multiplayer games using the Godot Engine 4.0 Network API.

Firstly, we are going to understand some fundamental aspects of how computers communicate through a network and the main protocols, including which ones are more relevant for making online multiplayer games.

After that, we will understand how Godot Engine 4.0 uses and provides both low- and high-level implementations for networking using its network API. We'll understand some core classes that we can use to pass data around to multiple computers on the same network. And then we'll focus on the high-level API known as `ENetMultiplayerPeer`.

With the fundamentals in place, we'll use the knowledge we just learned to turn local gameplay features into online gameplay features. To do that, we will develop five game projects:

- An online quiz game
- Checkers
- Pong
- A co-op platformer
- A top-down adventure

Then, we'll learn some techniques we can use to improve our players' experience by optimizing how their game sends, receives, and processes network data. We'll understand that we don't need constant updates and that we can do most of the gameplay with small bits of data and let the clients' computers fill the gaps on their own.

Throughout each chapter, you're going to do a role play of a network engineer working for a fictional, independent game development studio. In each chapter, you will apply your recently learned knowledge to a fictional problem presented by your studio's peers. You'll focus on the network aspect of each project they present, so you don't waste your precious time trying to understand unnecessary aspects.

In this chapter, you are going to learn the most important aspect of establishing a network of computers: to connect them all together. You'll see how this process happens, the reason for doing this, what's required to establish this connection, and how we can do that using the API that Godot Engine provides.

We will cover the following topics in this chapter:

- Introduction to a network
- Understanding the Godot Engine Network API
- Setting up the client side
- Setting up the server side
- Making your first handshake

By the end of the chapter, you'll have a client and server version of an application that establishes the connection of two or more computers. This is the very core of everything that we are going to see throughout the book, and with that knowledge, you'll understand how you can start making computers communicate within a network, which is exactly what you need to do in online multiplayer games.

Technical requirements

Godot Engine has its own standalone text editor, which is what we are going to use to code all our practical lessons.

As mentioned earlier, in this book you will do a role play of a network engineer of a fictional indie game studio. So, we will provide pre-made projects with all the non-network-related work ready. You can find them in the book's GitHub repository: `https://github.com/PacktPublishing/The-Essential-Guide-to-Creating-Multiplayer-Games-with-Godot-4.0`.

With the project properly added to your Godot Engine's project manager, open the project and move on to the `res://01.setting-up-a-server` folder. Here, you'll find what you need to follow this chapter (later part).

Introduction to a network

Making a network of connected computers is quite a task. In this chapter, we'll understand the core concepts of online networks. We'll also cover how Godot Engine provides solutions to each of the problems we may face in our quest to make online multiplayer games.

A **network** is a collection of interconnected devices that communicate with each other. In this communication, these devices exchange information and share resources with each other. You can have a local network, such as in a house or office, or a global network, such as the internet. The idea is the same.

For these devices to communicate they need to perform what we call a **handshake**. A handshake is how one device recognizes another device and establishes their communication protocols. This way, they know what they can request, what they expect to get, and what they need to send to one another.

A handshake begins with one device sending a message to another device. We call this message a *handshake request*. The devices use this message to start the handshake process. The one that sent the request waits for a message from the one that received it. We call this second message a *handshake response*.

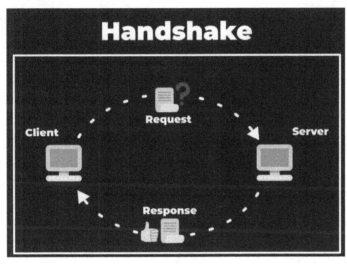

Figure 1.1 – The handshake procedure

When the requested device sends the confirmation through the handshake response, they establish their communication. After that, the devices start exchanging data. This wraps up the handshake process. We usually call the device that requests data the *client*. As for the one that provides data, we call it the *server*.

Note that we use these names for the very first interaction. After this first interaction, it is common that these roles change. In that sense, communication is *dynamic*. The server may request data from the client, and the client may provide data to the server.

In this chapter, we are going to make our first handshake using the Godot Engine Network API. We'll also create and synchronize players' data across the network. So, hold tight, as you'll learn the following:

- What the ENet library is and why we use it for games
- How we can make a handshake using the `ENetMultiplayerPeer` class

For that, you'll create a Godot project that lists connected players and allows them to edit and sync a line of text. It's a simple but elegant project that covers the basics of setting up an online multiplayer environment in Godot Engine.

Understanding the ENetMultiplayerPeer class

One of the Godot Engine Network API's core features is the `ENetMultiplayerPeer` class. By using this class, we can perform a handshake between our game server and clients.

The `ENetMultiplayerPeer` class is a high-level implementation of the ENet library. Let's understand this library and why we use it in online multiplayer games.

What is the ENet library?

ENet is a lightweight, open source networking library that is widely used in the game development industry. It is designed to be a high-performance, reliable, and easy-to-use library for creating multiplayer games and other networked applications. One advantage of the ENet library is that it's cross-platform and written in **C**. So, it's efficient with a small footprint and low overhead.

The library provides a simple and easy-to-use API that makes it easy for developers to create and manage network connections, send and receive packets, and handle network events such as disconnections and packet loss.

Packets, in this context, are small units of data that servers and clients transmit over the network. We use them to transmit information such as game state, player input, and other types of data between different devices on the network.

The ENet library offers support for multiple channels that allow us to easily create multiple streams of data, such as voice and video, within a single connection. This is excellent for many multiplayer games.

Another reason to use ENet in multiplayer games is its easy-to-use networking library that is based on the UDP protocol. This is a good chance to understand one of the main network protocols, so let's do it.

What is the UDP protocol?

The **UDP protocol** is a connectionless protocol that is well suited for real-time, high-bandwidth applications such as online gaming. This is because it has low latency and is able to handle high throughput. Just so we are on the same page, in the world of network terms, latency refers to the time between the transmission and receiving of data through the network.

For instance, it's very common to talk about lag in online multiplayer games: the time between the player performing an action and the game reacting to it. The next figure illustrates how latency works and is calculated:

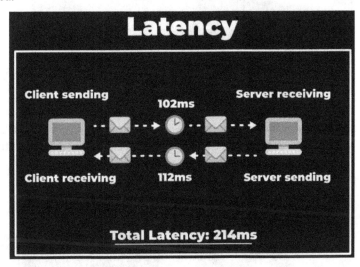

Figure 1.2 – Visual demonstration of latency

It's basically how much time it takes for data to cross the network, be properly handled by the server, and to provide a response to the client.

Throughput refers to how much data we can send through a given network route within a time period before it gets overwhelmed. For instance, this is a fundamental concept when we talk about **DDoS attacks**, where hackers overwhelm the server with an immense number of unsolved requests, preventing other clients from accessing the service. In the following figure, you can see a visual representation of the throughput concept:

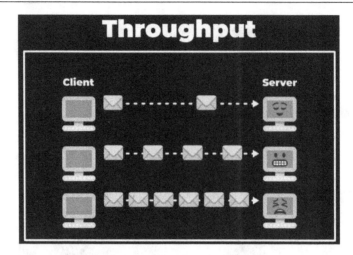

Figure 1.3 – Visual demonstration of throughput

The bandwidth is how big the available channel of communication in the network is. You can think of it as a pipe that streams data. A bigger pipe allows a lot of data, and big data, to be transmitted at any given time, while a small pipe may not even allow any data, of any size, to be transmitted. You can see this concept illustrated in the following figure:

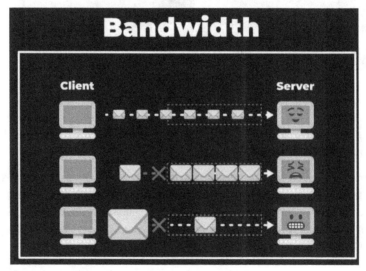

Figure 1.4 – Visual demonstration of bandwidth

Unlike the more commonly used **Transmission Control Protocol (TCP)**, **User Datagram Protocol (UDP)** does not establish a dedicated connection between two devices before transmitting data. Instead, it simply sends packets of data to a specified destination address without ensuring that the packets have been received or acknowledged.

Sounds… bad, right? But it's quite the opposite.

This lack of reliability is often seen as a drawback of UDP, but in the context of online multiplayer games, it can actually be an advantage. In games, where responsiveness and low latency are critical, the overhead of establishing and maintaining a connection can be a significant bottleneck.

By not requiring a dedicated connection, UDP allows for faster and more efficient transmission of data. Additionally, since UDP does not require the receiver to acknowledge receipt of packets, it is less affected by network congestion or delays, which can be critical for maintaining a stable and responsive connection in a high-bandwidth, high-latency environment such as online gaming.

Furthermore, the lack of reliability of UDP can actually be beneficial in the context of online multiplayer games. In games, where a small amount of packet loss or delay can have a large impact on the players' experience, it's important that the game can adapt to these types of network conditions. By not providing guarantees on packet delivery, UDP allows the game to handle packet loss and delay in a way that is most appropriate for the specific game and its mechanics.

Think about the following situation.

We establish a connection. In this connection, we update all players in the network about all other player avatars' positions in the world. This way, everyone shares the same world state.

If we use a TCP protocol, everyone will have to wait for every other player to send their position and confirm that they have received every change in every other player's position, while also trying to maintain the correct chronological order in which the positions have changed.

So, in this example, if a player moves five units to the left and sends 15 packets with all the movement data, including being idle, all other players must confirm that they have received all those 15 packets.

Using UDP, players can ignore every update but the latest one, which is the only relevant piece of information in real-time experience: what is the game-world state *now*? It doesn't matter how it gets to this point; it only matters that it is there at this very moment.

We are going to see that this causes some trouble as well. But we can create methods and understand techniques to mitigate those issues. We are going to talk about that in further chapters.

How does this connection happen?

To establish a UDP connection, we need two core things:

- The IP address of the peers, mainly the server

- The port over which they will exchange data

For test purposes, on all our projects we are going to use the `localhost` IP address. This is a shortcut to your local IP address mask. An IP address is like a house or apartment address. It is the exact location to which a given packet should be delivered and represents the address of the computer in the network. A port is essentially a specific channel in which the host allows a given communication to be established; we'll use the `9999` as our default port. There's nothing special about this one; it's just an arbitrary pick.

With this in mind, let's see for the first time the `ENetMultiplayerPeer` class in action. As you can imagine, this setup requires a two-sided approach. We need to set up a game architecture for our server and a different architecture for our client.

Let's start with the server architecture.

Creating the server

The `ENetMultiplayerPeer` class in the Godot Engine provides a convenient way to create and manage network connections for online multiplayer games. One of the most important methods of this class is the `create_server()` method, which is used to create a new server that can accept connections from clients. This method is simple to use and, besides having five arguments, it only requires one to get started:

- The first argument of the `ENetMultiplayerPeer.create_server()` method is the port on which the server will listen for incoming connections. This is the port number that clients will use to connect to the server. For example, if you want the server to listen on port `9999`, you would call `ENetMultiplayerPeer.create_server(9999)`. This is the only mandatory argument to call this method.

- The second argument is `max_clients`, which is the maximum number of clients that the server will allow to connect at the same time. This argument is optional, and if not specified, the server will allow up to 4,095 clients to connect.

- The third argument is `max_channels`, which is the maximum number of channels we allow the server to use per client. Channels are used to separate different types of data, such as voice and video, and are useful for creating multiple streams of data within a single connection. This argument is optional, and if not specified, the server will allow an unlimited number of channels.

- The fourth argument is `in_bandwidth`, which is the maximum incoming bandwidth that the server will allow per client. This argument is optional, and if not specified, the server will allow unlimited incoming bandwidth.

- The fifth argument is `out_bandwidth`, which is the maximum outgoing bandwidth that the server will allow per client. This argument is optional, and if not specified, the server will allow unlimited outgoing bandwidth.

Let's create our server in Godot Engine. Open up the project provided in the GitHub link given previously. After opening the project, execute the following steps:

1. Create a new scene and use a `Node` instance as the root.

2. Attach a new **GDScript** to this node, and name it `Server.gd`.

3. Save the scene and open the script.

4. Define a constant called `PORT` and set it to our default port number so the server can listen to it:

   ```
   const PORT = 9999
   ```

5. Create a new `ENetMultiplayerPeer` using the `new()` constructor. Let's store it in a variable called `peer`:

   ```
   var peer = ENetMultiplayerPeer.new()
   ```

6. In the `_ready()` function, call the `create_server()` method on the `peer` variable, passing in the `PORT` constant as an argument:

   ```
   func _ready():
       peer.create_server(PORT)
   ```

7. Still in the `_ready()` callback, assign the `peer` variable to the built-in `multiplayer` member variable of this node:

   ```
   multiplayer.multiplayer_peer = peer
   ```

8. Connect the `peer_connected` signal of the `multiplayer` variable to a function called `_on_peer_connected`. We'll create this callback method next:

   ```
   multiplayer.peer_connected.connect(_on_peer_connected)
   ```

9. Create a new method called `_on_peer_connected()`, which should receive `peer_id` as the argument:

   ```
   func _on_peer_connected(peer_id):
   ```

10. In the _on_peer_connected() function, use print() to print the passed peer_id argument on the console:

```
print(peer_id)
```

The complete script should look like this:

```
extends Node

const PORT = 9999

var peer = ENetMultiplayerPeer.new()

func _ready():
    var error = peer.create_server(PORT)
    multiplayer.multiplayer_peer = peer
    multiplayer.peer_connected.connect
        (_on_peer_connected)

func _on_peer_connected(peer_id):
    print(peer_id)
```

It's important to note that this script uses the built-in multiplayer member variable that every Node instance has on Godot Engine 4.0 Network API, which is an instance of the MultiplayerAPI class.

Done: we have our server ready. Told you it would be simple!

Creating the client

Next up, let's create our client. The process is quite similar. The major difference is that the client needs the server IP address to find it on the network.

We use the ENetMultiplayerPeer.create_client() method to connect a client to a server. This method is very simple to use and requires only two arguments to work:

- The first argument of the create_client() method is the address of the server. This can be either the server's IP or hostname. For instance, if you want the client to connect to a server with the IP address 192.168.1.1, you would call create_client("192.168.1.1"). But to make things simpler, we'll use "localhost", which is a shortcut to our own IP address mask.

- The second argument of the create_client() method is the port on which the server is listening for incoming connections. This is the port number that the client will use to connect to the server. For example, if the server is listening on port 9999, you would call create_client("192.168.1.1", 9999).

- The third argument of the `create_client()` method is `channel_count`, which is the number of channels that the client will use to communicate with the server. Channels are used to separate different types of data, such as voice and video, and are useful for creating multiple streams of data within a single connection. This argument is optional, and if not specified, the client will use a default value of 1 channel.

- The fourth argument of the `create_client()` method is `in_bandwidth`, which is the maximum incoming bandwidth that the client will allow per connection. This argument is optional, and if not specified, the client will use a default value of 0, allowing an unlimited incoming bandwidth.

- The fifth argument of the `create_client()` method is `out_bandwidth`, which is the maximum outgoing bandwidth that the client will allow per connection. This argument is optional, and if not specified, the client will use a default value of 0, allowing an unlimited outgoing bandwidth.

- The sixth argument of the `create_client()` method is `local_port`, which is the local port that the client will bind to. This argument is optional, and if not specified, the client will use a default value of 0.

Now, let's see how we can create the *client* side of this connection so it can connect with our *server* and establish their handshake:

1. Create a new scene and add a `Node` instance as the root.

2. Attach a new script to it.

3. Save the script as `Client.gd`.

4. In the script, define a constant called `ADDRESS` and set it to the server's IP. In this case, we are going to use `"localhost"`:

   ```
   const ADDRESS = "localhost"
   ```

5. Define a constant called `PORT` and set it to be our default port number. It's very important that this matches the number we used in `Server.gd`, otherwise these peers won't be able to find each other:

   ```
   const PORT = 9999
   ```

6. Create a new `ENetMultiplayerPeer` using the `new()` constructor and store it in a variable called `peer`:

   ```
   var peer = ENetMultiplayerPeer.new()
   ```

7. In the `_ready()` callback, call the `create_client()` method on the `peer` variable, passing in the `ADDRESS` and `PORT` constants as arguments:

```
func _ready():
  peer.create_client(ADDRESS, PORT)
```

8. Assign the `peer` variable to the built-in `multiplayer` member variable of the node:

```
multiplayer.multiplayer_peer = peer
```

The complete script should look like this:

```
extends Node

const ADDRESS = "localhost"
const PORT = 9999

var peer = ENetMultiplayerPeer.new()

func _ready():
    peer.create_client(ADDRESS, PORT)
    multiplayer.multiplayer_peer = peer
```

Alright, we have our server and our client ready. Now, how do we test them?

Testing our handshake

Godot Engine 4.0 has a useful feature for debugging: the ability to open multiple independent instances of the game. This feature allows us to test different scenes at the same time, making the debugging process much easier and faster.

To open multiple instances of the game, we need to select one option from up to four options in the **Debug | Run Multiple Instances** menu.

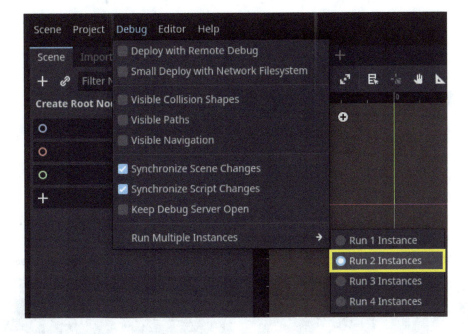

Figure 1.5 – The Run Multiple Instances menu

Then, as soon as we press the **Run Project** or **Run Current Scene** button, Godot will launch the instances we've set previously. Let's stick with two instances for this project.

This feature is incredibly useful for testing online multiplayer games, as it allows us to open a server and a client in the same run. But, as you can see, it's not very straightforward. When we run the project, it actually opens two instances of the same scene.

Let's create a minimal menu where we can select whether we are a client or a server:

1. Create a new scene and use `Control` as the root and name it `MainMenu`.

2. Add a `Label` node as a child of the root node.

3. Add two `Button` nodes as children of the root node.

4. Give the first `Button` the name `ClientButton` and the second one `ServerButton`:

Figure 1.6 – MainMenu's Scene tree structure

5. Set the `Button` nodes' `text` properties to **I'm a client** and **I'm a server** respectively and position them side by side in the middle of the screen.

6. Set the `Label` node's `text` property to **Are you a…** and position it in the middle of the screen.

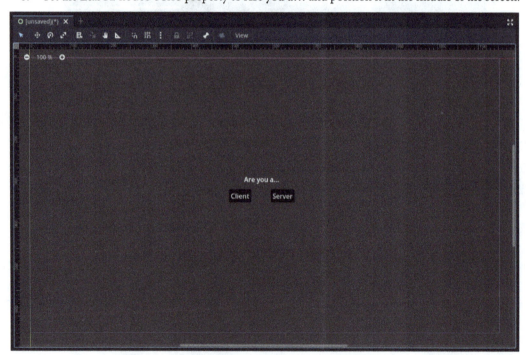

Figure 1.7 – MainMenu's scene UI

7. Attach a new **GDScript** instance to the `MainMenu` node and open it.

8. Connect the **ClientButton**'s pressed signal to a function called `_on_client_button_pressed`.

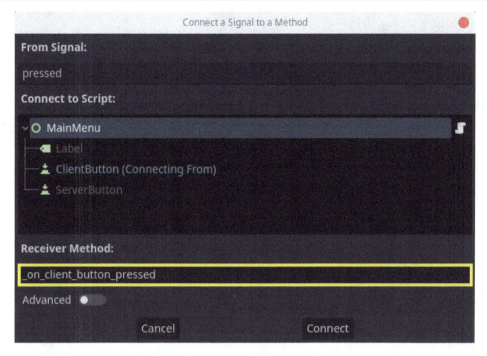

Figure 1.8 – ClientButton's pressed signal connection

9. Connect the pressed signal of the **ServerButton** to a function called `_on_server_button_pressed`.

10. In the `_on_client_button_pressed()` callback, let's call the `change_scene_to_file()` method on the `get_tree()` instance, passing in `"res://Client.tscn"` as the argument:

```
extends Control

func _on_client_pressed():
    get_tree().change_scene_to_file
        ("res://Client.tscn")
```

11. In the `_on_server_button_pressed()` callback, do the same as before, passing `"res://Server.tscn"` instead.

The complete script should look like this:

```
extends Control

func _on_client_pressed():
    get_tree().change_scene_to_file
```

```
        ("res://Client.tscn")

func _on_server_pressed():
    get_tree().change_scene_to_file("res://Server.tscn")
```

Now, let's make sure we save the scene before we test it. After that, all we need to do is hit the **Run Current Scene** button and watch the scene come to life. All the hard work has been done, and now all that's left is to appreciate the results.

Once we have the two debug instances running, we need to pick one to be the server first. For that, we can press **ServerButton**. This will launch our `Server.tscn` scene and start listening for incoming connections.

Then, in the other instance, we need to press **ClientButton**. This will launch the `Client.tscn` scene and try to connect to the server. If everything goes as expected, we should get `peer_id` printed in the console of the server instance.

This means that the client and the server have successfully established a connection and are now ready to start exchanging messages. Congratulations, you've just created your first handshake!

Summary

In this chapter, we went through the fundamentals of network connections, which is to establish the connection through a procedure known as the handshake.

The handshake ensures that two computers recognize each other in a network and establish the protocols of this communication. This is important to understand, as it is the core of all our further endeavors. Without this at our disposal, our players and our server would be disconnected. One would be sending data to the void, while the other would be infinitely waiting for something to arrive.

Talking about sending data, now that we have our computers connected and open to receive and send data back and forth, it's time to see how to do that. Throughout this chapter, you saw how you can properly establish a connection using the ENet library and how Godot Engine provides a high-level approach to handshaking, to the point that we can barely see if there was an actual handshake or not.

In the next chapter, we are going to use the UDP protocol to also establish a connection between client and server. But this time, we are going to dig a bit further and actually send data both from the client to the server and the other way around.

It's important to use the UDP protocol to understand what might be happening under the hood when we finally start to get used to the Godot Engine `ENetMultiplayer` API.

Now let's see the dirty and messy world of low-level data transmission in the next chapter, so we can understand later how much easier our lives are made with the new high-level network API!

2
Sending and Receiving Data

In the previous chapter, we saw how we can establish a connection between two computers using the high-level Godot Engine `ENetMultiplayerPeer` API. But what do we do after that? Why do we establish connections between computers? The foundation of a network is the communication between the connected computers, allowing them to send and receive data. This data is transferred by breaking down the content into small chunks called **packets**.

Each packet is like a postcard containing the necessary information, such as the sender's and receiver's IP addresses, the communication port, and the message's content. We then send these packets over the network, where they can be routed to their intended recipient. Using communication protocols, such as the UDP protocol, we break the data into packets at the sending end and reassemble them at the receiving end of the relationship.

In this chapter, we will discuss the fundamentals of how packets are sent and received and what makes the UDP protocol unique. For that, we'll need to go a bit lower and use the Godot Engine `UDPServer` and `PacketPeerUDP` classes. These are lower-level API classes, so we will go through some intense content here.

We will cover the following topics in this chapter:

- Understanding packets
- Introduction to the **JavaScript Object Notation (JSON)** format
- Sending packets with `PacketPeerUDP`
- Listening to packets with `UDPServer`
- Authenticating the player
- Loading the player's avatar

Technical requirements

In this chapter, we're going to keep up with our project in Godot Engine, but this time, we are going to use the files provided in the `res://02.sending-and-receiving-data` folder. So, if you haven't already done so, download the project's repository using this link: `https://github.com/PacktPublishing/The-Essential-Guide-to-Creating-Multiplayer-Games-with-Godot-4.0`.

Then, with the project added to your Godot Engine project manager, open the project and proceed to the `res://02.sending-and-receiving-data` folder.

Understanding packets

Packets are fundamental building blocks of communication over the network using the UDP protocol. They are small chunks of data that contain all the necessary information to reach their intended recipient. This includes the sender's and receiver's IP addresses, the communication port, and the message's content.

Senders send packets to the receiver over the network. The receiving end reassembles the packets, allowing the receiver to understand the message sent. This process is known as **packet switching**. You can see a visual representation of this here:

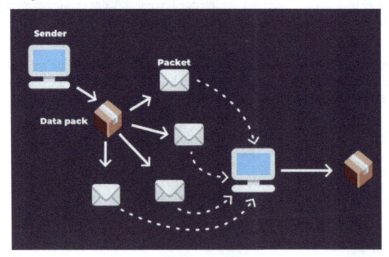

Figure 2.1 – Packet switching process

Unlike other protocols, such as the TCP protocol, the UDP protocol does not guarantee that packets will arrive in the same order as they were sent. This means that the protocol is less reliable but more efficient and faster.

UDP is also different from other protocols due to its lack of connection state. Each packet contains all the data it needs to reach its receiving end. We address them individually, and the network routes them based on each packet's own information. This contrasts with the TCP protocol as the latter needs to set up a prearranged, fixed data channel through a traditional handshake procedure.

This means that we can send these packets using the UDP protocol without a handshake. As long as our server is listening to messages at the specific port we've assigned, it will be able to receive the sender's message.

Due to all that, the UDP protocol is more efficient for sending gameplay data across a network because it's fast and doesn't need to wait for confirmation of each packet in order at the receiver's end. This is a huge advantage for online multiplayer games, especially ones where the player's reaction time is important for the gameplay.

It's also common to use the UDP protocol for quick message systems and even voice calls. One issue that may come with using UDP for voice calls is that sometimes the audio doesn't reach the other side in the correct order, or in any order at all. This causes some issues, but since the communication is meant to be real time and users can ask the person at the other end of the conversation to repeat, the UDP protocol has become the go-to solution for this type of service. And this is what is important to understand—when it is and when it isn't the proper choice.

Now that we have taken a glance at the protocols in which we can exchange data across the network, we need to understand what this data looks like. Can we send instances of objects across the network? How will they assemble at the receiver's end?

Network communication is a bit lower level in this sense; we need to send only relevant information in data structures that both the sender and the receiver ends can understand. And for that, we commonly avoid passing binary data around, such as objects.

Instead, we serialize the important information and transmit the necessary chunks across the network so that the receiver end can create a whole new clone of the object using only the data we transmitted. This is way more reliable and allows for smaller bandwidth usage. One common data structure that we use is dictionaries in the format of JSON files.

Introduction to the JSON format

In network programming, transmitting objects directly through the network is not always reliable, as the data may get corrupted or lost in transit. Moreover, transmitting objects containing executable code may pose a security risk if the code is malicious. That's why it's a common practice to use data serialization to convert objects into a format that can be easily transmitted over the network.

One of the most commonly used data serialization formats is JSON. JSON is a lightweight, text-based format that can represent complex data structures such as arrays and objects, making it an ideal choice for network communication.

When using the Godot Engine network API with UDP, sending and receiving JSON files is a common practice. With JSON, we can serialize and deserialize data quickly and efficiently. JSON files are human-readable, making it easier for developers to debug and troubleshoot issues. JSON files are also flexible, meaning we can cherry-pick only the relevant data we need to send, making network communication even more efficient.

Unlike binary formats, JSON files are easy to read and modify as well. This makes it easier to debug and troubleshoot any issues that may arise during the transmission of data.

Now that we understand the advantages and the overall idea behind the JSON format, how do we use it properly? How does a JSON file help us transmit data around a network and keep players in the same game context?

As mentioned in this section, serialization is how we cherry-pick only the necessary information about a data structure, such as an object, and translate it into a format that we can pass around, and even store, to reconstruct the previous data structure. Serialization is one of the most important skills to learn in software engineering fields, including networks.

It is through serialization that we can translate the state of our application so that other instances of our application can further replicate this state through time—for instance, to make a save and load system or through space, as we are going to do in online multiplayer games. So, let's understand how serialization works and how to do it.

Serialization

Serialization in a networking context refers to the process of converting a complex data object, such as a Sprite2D node, into a simple, linear representation that we can store in a file. For instance, *.tscn files are serialized files that represent a scene in Godot Engine's editor.

Serialization involves converting an object into a format that can be easily reconstructed on another machine or in another context. This can involve encoding the object's properties, data, and other relevant information in a standardized format, such as JSON. Serialization is essential in network communication because it allows data to be transmitted and received efficiently and reliably, while also enabling interoperability between different programming languages and systems.

For instance, if we want to recreate a Sprite2D node on the client's side based on data provided by the server, we can serialize important properties such as its position, rotation, scale, and texture. It would look like this:

```
{
    "position": {
        "x": 2244,
        "y": 1667
    },
    "rotation": 45,
```

```
  "scale": {
    "x": 2,
    "y": 2
  },
  "texture_path": "res://assets/objects/Bullet.png"
}
```

So, on the client's side, we instantiate a new `Sprite2D` node and use this data to ensure it represents what the server wants the client to see. We are going to use serialization a lot moving forward. In Godot, we have the `JSON` helper class for creating and parsing JSON data.

The `JSON.stringify()` method is used to serialize an object or a data type, such as an integer or a dictionary, into a JSON-formatted string. This method takes an object as input and returns a string containing the JSON representation of the input object.

The string can then be transmitted over the network, stored in a file, or used in any other context where a string representation of the object is needed. The resulting string can easily be deserialized back into an object using the `JSON.parse_string()` method.

On the other hand, the `JSON.parse_string()` method is used to deserialize a JSON-formatted string back into a recognized Godot data type or object. This method takes a string as input and returns the deserialized data. The resulting object can then be used in any context where the original object was needed.

When deserializing the JSON string, the method takes care of mapping the JSON values to the appropriate Godot Engine data types. This includes mapping strings to strings, numbers to numbers, and Booleans to Booleans, as well as parsing more complex types such as dictionaries and objects.

With both `JSON.stringfy()` and `JSON.parse_string()` methods, Godot Engine provides a simple and reliable way to convert data into a format that can be transmitted over the network or stored in a file.

We saw how we can translate our relevant data into an understandable standard format that we can store, transmit, and recreate at the receiver end. Let's understand how we can pass this data around in the network.

This is fundamental knowledge when we deal with online multiplayer games because it's through this process that we will be able to recreate objects and even the whole game state across players, making them share the same game world.

Sending packets with PacketPeerUDP

Now, let's move on to practical knowledge. In this chapter, your task is to implement a login system for a game. Our project already has a cool user interface and is able to gather player data, such as their login name and password. Your mission is to make sure that only authorized players can access the game's content by implementing a secure authentication feature.

Once a player successfully logs in, you need to display their character's avatar based on what we have saved in our database. As a network engineer, you understand the importance of security when it comes to online systems. You know that a robust authentication system is essential to ensure that only legitimate users are granted access to the game's content.

Therefore, you will need to develop a login system that checks players' credentials against a secure database and verifies if they have permission to access the game's features or not.

With your skills and experience, you need to create a system that will provide an excellent user experience while keeping players' data secure. So, take up the challenge, and let's create a login system that will be a testament to your skills as a network engineer!

In our project repository, open the `res://02.sending-and-receiving-data//MainMenu.tscn` scene, and let's get started.

Creating an AuthenticationCredentials Autoload

In Godot Engine, **Autoloads** are singletons that Godot loads automatically when the game starts. We can create and edit them in the editor itself and access them from any script in the game. We use Autoloads to store game-wide data or to provide global functionality, making them a convenient way to carry players' credentials across the game.

One of the main advantages of using Autoloads for carrying player credentials is that they are available throughout scene changes. This means that any script in the game can access the Autoload and retrieve the players' credentials when needed. This eliminates the need to pass credentials from one script to another, making the code cleaner and easier to maintain.

In addition to that, since Autoloads are persistent throughout the game's lifetime, as long as players don't close the game, we can access their credentials.

This can make the process of implementing a login system with authentication features more efficient and streamlined.

So, let's create our `AuthenticationCredentials` Autoload, as follows:

1. Create a new scene using a `Node` node as the root node.

2. Rename the root node `AuthenticationCredentials`.

3. Attach a new script to it, save it as `AuthenticationCredentials.gd`, and open it.

4. Create a variable to store the player's username; we can name this variable `user`, and it should be an empty string by default:

    ```
    extends Node
    var user = ""
    ```

5. Then, create a variable to store the session's token when we successfully validate a login:

    ```
    var session_token = ""
    ```

6. Save the scene, then go to **Project** | **Project Settings** and open the **Autoload** tab.

7. In the **Path** field, click on the small folder icon:

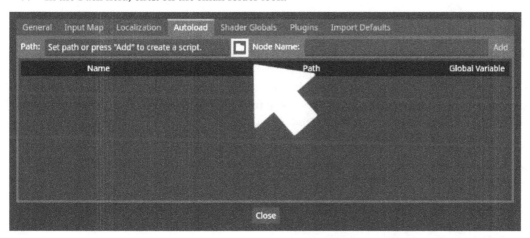

Figure 2.2 – Autoload tab in the Project Settings menu

8.   From the pop-up menu, select `AuthenticationCredentials.tscn`:

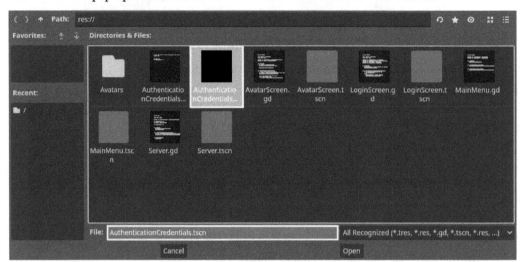

Figure 2.3 – Selecting the AuthenticationCredentials scene from the File menu

9.   Leave the **Node Name** field as **AuthenticationCredentials** and click on the **Add** button.

And there we have it. Now, you can access the variables and functions defined in the script of the `AuthenticationCredentials.gd` scene from anywhere in your project by calling the `AuthenticationCredentials` singleton.

This is useful for keeping track of global state across multiple scenes and nodes in your game. It's important to note that this Autoload should only exist on the client's side of a multiplayer game, and not on the server's side. So, make sure to remove it from your server application.

Now, let's see how we can gather and send players' credentials to the server. For that, we are going to work on the very login screen itself! Open `LoginScreen.tscn` and let's move on to the cool stuff.

Sending players' credentials

The **LoginScreen** scene is the gateway to the game world. It's essentially a `Control` node named **LoginScreen** with a user interface that captures players' credentials so that we can authenticate them and give them access to our world:

Figure 2.4 – The LoginScreen scene's node hierarchy

The interface includes two `LineEdit` nodes, one called **UserLineEdit** and another one called **PasswordLineEdit**. These `LineEdit` nodes allow players to input their login credentials. In case of any errors, we can use the **ErrorLabel** node to display any necessary messages.

As we collect the players' credentials here, we can use the **LoginButton** node to trigger the login procedure. With this scene in place, our players can securely access their avatar screen once they successfully log in.

But now, we need to work on validating their logins before loading their avatar. So, let's get our hands dirty. Proceed as follows:

1. Open the `LoginScreen.gd` script and go to the `send_credentials()` function.

2. Inside the `send_credentials()` function, create a dictionary called `message` that contains the user credentials we will authenticate in the server.

3. To store these credentials, create a key in the message dictionary called `'authenticate_credentials'`; its value should also be a dictionary. We'll use it to store players' credentials.

4. Use the `user_line_edit` and `password_line_edit` text properties to capture the player's input for their username and password, respectively:

    ```
    var message = {'authenticate_credentials':
        {'user': user_line_edit.text, 'password':
            password_line_edit.text}}
    ```

5. Instantiate a new `PacketPeerUDP` object called `packet` using the `PacketPeerUDP.new()` constructor:

    ```
    var packet = PacketPeerUDP.new()
    ```

6. Connect the `packet` object to the server's address and port using the `connect_to_host()` method. Here, we are using our default `ADDRESS` and `PORT` constants that represent the IP address and port number of the server to which the client is connecting. They are `127.0.0.1` and `9999` respectively:

```
packet.connect_to_host(ADDRESS, PORT)
```

7. Serialize the message dictionary object into a JSON-formatted string using the `JSON.stringify()` method, and send it to the server using the `packet.put_var()` method:

```
packet.put_var(JSON.stringify(message))
```

8. Create a `while` loop to wait for a response from the server. The `packet.wait()` method waits for a packet to arrive at the bound address. It returns an `OK` error constant if it receives a packet; otherwise, it returns an error code based on Godot's error constants. So, we can use that to wait for the arrival of our packet at the server's end:

```
while packet.wait() == OK:
```

9. When we receive a response, we need to deserialize the response data from JSON format back into a dictionary object using the `JSON.parse_string()` method. Let's store that in a variable called `response`:

```
var response = JSON.parse_string
    (packet.get_var())
```

10. Check the `response` dictionary for the presence of an authentication token using the `in` operator. If the `"token"` string is present, store its value in `AuthenticationCredentials.session_token`:

```
if "token" in response:
    AuthenticationCredentials.session_token =
        response['token']
```

11. After that, we can also store the `user` present in the message we got from the server as our player's username:

```
AuthenticationCredentials.user = message
    ['authenticate_credentials']['user']
```

12. Update the user interface to indicate a successful authentication, and switch to the `AvatarScreen.tscn` scene. If the token is not present, display an error message to the player:

```
error_label.text = "logged!!"
```

13. Then, after all that, we can change the scene to the actual avatar screen using the `get_tree().change_scene_to_file("res://AvatarScreen.tscn")` method and break the `while` loop:

```
get_tree().change_scene_to_file
    ("res://AvatarScreen.tscn")
break
```

14. In case we get a response from the server, and it doesn't have the `"token"` key in it, we display an authentication failed message using the `error_label.text` and also break the `while` loop:

```
else:
    error_label.text = "login failed,
        check your credentials"
    break
```

At this point, the `send_credentials()` method should look like this:

```
func send_credentials():
    var message = {'authenticate_credentials':
        {'user': user_line_edit.text, 'password':
            password_line_edit.text}}

    var packet = PacketPeerUDP.new()
    packet.connect_to_host(ADDRESS, PORT)
    packet.put_var(JSON.stringify(message))

    while packet.wait() == OK:
        var data = JSON.parse_string(packet.get_var())
        if "token" in data:
            error_label.text = "logged!!"
            AuthenticationCredentials.user = message
                ['authenticate_credentials']['user']
            AuthenticationCredentials.session_token =
                data['token']
            get_tree().change_scene_to_file
                ("res://AvatarScreen.tscn")
            break
        else:
            error_label.text = "login failed,
                check your credentials"
            break
```

Now that we've seen how the client side works and what it will do with players' data, let's understand how the other side of this connection will receive this data and handle it. For that, open the `Server.tscn` scene.

Listening to packets with UDPServer

Welcome to our Godot Engine server scene! This scene is where our game's server logic is implemented.

The server is the backbone of our game, responsible for authenticating players and providing them with data about their avatars, such as their name and texture file. This node is called `Server`, and it has a pre-written script that includes some essential variables. Among them are two vital variables: `database_file_path` and `logged_users`.

The `database_file_path` variable is the path to the `FakeDatabase` JSON file, which represents a fake database that holds the players' data. The `logged_users` variable is a dictionary that stores players who are currently logged in.

These variables are crucial to our server's functionality, and we will use them to authenticate players and provide them with the data they need.

Let's implement the `Server` node's most important feature, which is to listen to packets. Proceed as follows:

1. Open the `Server.gd` file.

2. Declare a `server` variable and set it to `UDPServer.new()`. This creates a new instance of the `UDPServer` class that will allow us to listen to incoming connections:

    ```
    var server = UDPServer.new()
    ```

3. In the `_ready()` function, call the `listen()` method on the `server` variable, passing our default `PORT` constant as an argument. This will start the server and make it listen for incoming connections:

    ```
    func _ready():
        server.listen(PORT)
    ```

4. In the `_process(delta)` function, call the `poll()` method on the `server` variable to check for any incoming messages. This method will not block the game loop, so we can call it in the `_process(delta)` function safely:

    ```
    func _process(delta):
        server.poll()
    ```

5. Call the `is_connection_available()` method on the `server` variable to check whether a client sent a message. If it returns `true`, call the `take_connection()` method to obtain a `PacketPeerUDP` instance that we can use to read the incoming message:

```
if server.is_connection_available():
    var peer = server.take_connection()
```

6. Use the `get_var()` method on the `PacketPeerUDP` instance we get to obtain the incoming message. Since we know that the message is a string in JSON format, we can use the `JSON.parse_string()` method to convert it to a dictionary object that we can work with:

```
var message = JSON.parse_string(peer.get_var())
```

7. Check whether the incoming message contains the `"authenticate_credentials"` key. If it does, call the `authenticate_player()` function, passing `peer` and `message` as arguments:

```
if "authenticate_credentials" in message:
    authenticate_player(peer, message)
```

We will create the `authenticate_player()` method in a moment, but for now, our script should look like this:

```
extends Node
const PORT = 9999
@export var database_file_path =
    "res://FakeDatabase.json"

var database = {}
var logged_users = {}
var server = UDPServer.new()

func _ready():
    server.listen(PORT)

func _process(delta):
    server.poll()
    if server.is_connection_available():
        var peer = server.take_connection()
        var message = JSON.parse_string
            (peer.get_var())
        if "authenticate_credentials" in message:
            authenticate_player(peer, message)
```

We just saw how we can open a communication channel between the client and the server and start to listen to messages. With that, we can filter these messages so that the server knows what the client is requesting—in our case, to authenticate the players' credentials.

This is a low-level implementation of a network API. With that, we can create standard message formats and contents that trigger events on the server side and expect standard responses from the server. Let's see how our server replies to this client request.

Authenticating the player

Authenticating player credentials is a crucial aspect of any multiplayer game. In our project, we are building a login system for a game using Godot Engine. The login system allows players to log in with their username and password and then displays their character's avatar upon successful login.

We are going to use a fake database, stored as a JSON file, to represent the players' credentials. While this approach is simpler than using a full-fledged database management system, it has its own security risks. So, be aware of the risks of this approach in a production-ready project.

To authenticate player credentials in our project, we will also use Godot's `FileAccess` class to load the fake database from the JSON file and parse the data. This will allow us to compare players' login credentials with the data in the database and authenticate the player if the credentials match.

Loading a fake database

Now, let's load our database so that we can check whether the data we got from the player's client matches anything on our server. In a nutshell, a database is an organized collection of data. In our case, we'll use a JSON file format as our database.

The advantage of using JSON files as databases is that they are easy to manipulate, and you don't need to have prior knowledge of database structures and safety.

For instance, our fake database consists of the following:

```
{
  "user1": {
    "password":"test",
    "avatar":"res://Avatars/adventurer_idle.png",
    "name":"Sakaki"
  },
  "user2": {
    "password":"test",
    "avatar":"res://Avatars/player_idle.png",
    "name":"Keyaki"
  }
}
```

You can even open it in the very Godot Text Editor itself; just double-click the `res://FakeDatabase.json` file provided in our base project.

The preceding JSON file represents a simple database that contains two user entries, `"user1"` and `"user2"`, each with a corresponding set of data. The data contained for each user includes a password, an avatar, and a name.

The `"password"` field holds the plain-text password for each user. This is a very simple approach to storing passwords, as it is not secure due to the possibility of it being compromised. However, it is suitable for educational purposes.

The `"avatar"` field contains a reference to a file that represents the user's avatar. In this case, it is referencing two different image files from our game, one for each user.

Finally, the `"name"` field simply holds a string that represents the player's avatar name.

Note that a database file shouldn't by any means be available to the client. So, in your final project, make sure to remove your database file from Godot's project and into a safe database device.

While JSON files are a great choice for certain projects, they may not be suitable for others. Here are some pros and cons to consider:

- **Pros**:

    - They are easy to read and write, making them a great choice for small projects or when the speed of development is a priority

    - JSON files can be parsed natively by most programming languages, including GDScript as we saw previously, which means you don't need to install any additional software or libraries to work with them

    - As we just saw, JSON files are human-readable and can be opened and edited using a simple text editor, which makes them great for debugging

- **Cons**:

    - They don't scale well for large projects with many concurrent users, as there may be issues with data consistency and performance

    - JSON files are not as flexible as other database formats when it comes to querying data and performing complex operations

To load and read the data from our JSON fake database file, we'll use Godot Engine's `FileAccess` class.

The `FileAccess` class is a built-in Godot class that provides an interface to load, read, write, and save files to and from the user's disk. It is a powerful tool that is essential for any game or application that needs to access files from the user's device.

Let's dive into the specifics of how to use this class to load and parse our JSON fake database file into our game, as follows:

1. Go to the `load_database()` function in the `Server.gd` script.

2. In the function, create a new instance of the `FileAccess` class by calling the `open` method and passing in the path to the JSON file as the first argument and `FileAccess.READ` as the second argument. The `READ` constant tells the `FileAccess` class that the file should be opened for reading:

```
func load_database(path_to_database_file):
    var file = FileAccess.open(path_to_database_file,
        FileAccess.READ)
```

3. Once the file is open, call the `get_as_text()` method to read the contents of the file as a text string:

```
var file_content = file.get_as_text()
```

4. Next, parse the contents of the file as a JSON string using the `JSON.parse_string()` method and store the resulting dictionary in the `fake_database` variable:

```
fake_database = JSON.parse_string(file_content)
```

Before we move on to replying to the player's authentication request, let's see how this function looks at the end of these steps:

```
func load_database(path_to_database_file):
    var file = FileAccess.open(database_file_path,
        FileAccess.READ)
    var file_content = file.get_as_text()
    database = JSON.parse_string(file_content)
```

With our database in place, we can look at our valid players and check whether the credentials we receive in the message sent by the client match the credentials we have stored. Ideally, we would use a safer format to avoid any data leaks or hack attacks, but this should do for our small application.

Now, let's see how we can reply to the client with a valid response based on whether the player was successfully authenticated or if the authentication failed. In the former case, we will provide an authentication token to the player so that they can use it across their play session in order to keep them logged in without further authentication procedures.

Replying to the authentication request

When a client sends their credentials to the server to be authenticated, the server will receive them and start the authentication process. The server will use the credentials to search in our fake database, which contains user data, for a matching record. If the credentials match, the server will generate a session token and send it back to the client.

A session token is a unique string of characters that identifies the client on the server side, and the client must present it on all subsequent requests to the server to prove their identity.

To validate credentials, we call the `load_database` function, which we can do in the `_ready()` function to load the fake database into our server.

Then, we will use the username that the player provided through the **LoginScreen** node to look up the record in the dictionary. If the record exists, we will compare the password provided by the client with the one stored in the record. If they match, we will generate a session token and store it in the `logged_users` dictionary, along with the username, to keep track of the authenticated users.

If a client tries to use an invalid or expired session token, the server will deny the request, and the client will need to authenticate again. This way, we can ensure that only authenticated clients have access to the server's resources as they play.

Now, let's move on to the `authenticate_player()` function and create our authentication logic. Proceed as follows:

1. Access the `authenticate_credentials` key from the `message` dictionary and store it in the `credentials` variable, like so:

    ```
    func authenticate_player(peer, message):
        var credentials = message
            ['authenticate_credentials']
    ```

2. Check whether the `user` and `password` keys are present in the `credentials` dictionary by running the following code:

    ```
    if "user" in credentials and "password" in
        credentials:
    ```

3. If the keys are present, extract the values of the `user` and `password` keys from the `credentials` dictionary and store them in separate variables:

    ```
    var user = credentials["user"]
    var password = credentials["password"]
    ```

4. Check whether the `user` key we just stored is present in our `fake_database` dictionary keys:

    ```
    if user in fake_database.keys():
    ```

5. If the `user` is key present, check whether the `password` key matches the one stored in the `fake_database` dictionary:

```
if fake_database[user]["password"] == password:
```

6. If the `password` key matches, generate a random integer token and store it in the `logged_users` dictionary with `user` as the key so that we can always check them when necessary:

```
var token = randi()
logged_users[user] = token
```

7. Create a dictionary called `response` with a single key-value pair. The key is `token` and the value is the `token` variable:

```
var response = {"token":token}
```

8. Send the `response` dictionary back to the client in JSON format using the `peer.put_var()` method:

```
peer.put_var(JSON.stringify(response))
```

9. If the password does not match, send an empty string to the client to indicate that the authentication failed:

```
else:
    peer.put_var("")
```

With that, we should have a method that properly handles and replies to the player's authentication request. Let's see how it ended up:

```
func authenticate_player(peer, message):
    var credentials = message['authenticate_
        credentials']
    if "user" in credentials and "password" in
        credentials:
        var user = credentials["user"]
        var password = credentials["password"]
        if user in database.keys():
            if database[user]["password"] == password:
                var token = randi()
                var response = {"token":token}
                logged_users[user] = token
                peer.put_var(JSON.stringify(response))
            else:
                peer.put_var("")
```

Now, let's move on to one important part of this whole process. The player will get a request with a token that, as we saw in the *Sending Player's Credentials* section, they will store in the `AuthenticationCredentials` Autoload. So, after that, the player's game will change the scene to **AvatarScreen** and try to request their avatar.

Let's see how players will be able to keep their session valid throughout this process. The following section is fundamental even after the player actually starts playing the game. So, stay tuned to understand how we can always ensure the player is still holding a valid token.

Maintaining the player's session

One of the most important aspects of any online game is keeping the player's session alive throughout their playtime. In our project, we are going to make sure that the player's token is available throughout the whole game session, even when changing between different scenes. This way, we can maintain the player's identity as they play the game.

To achieve this, we will store the token on the player's machine using the `AuthenticationCredentials` singleton. This way, the player's token will be available to all the game's scripts, allowing us to check whether the player is still authenticated before proceeding to any other scene.

By keeping the token on the player's machine, we can avoid constant login requests to the server to authenticate the player again, ensuring faster and smoother gameplay. To ensure that the player's credentials are still valid, we will use the `get_authentication_token()` method to allow the player's client to make a request to the server for their authentication token.

We call this method whenever the player is about to transition to a new scene or when a certain amount of time has passed since their last request. This way, we can ensure that the player is still authenticated and can proceed with their gameplay without any issues.

So, still in the `Server.gd` script, go to the `get_authentication_token()` method, and let's start providing players with what they need to play our game moving on! Proceed as follows:

1. Inside the `get_authentication_token()` method, let's extract the user's information from the `message` argument. For that, we can create a new variable called `credentials` and assign it the value of the `message` argument:

   ```
   func get_authentication_token(peer, message):
       var credentials = message
   ```

2. Then, let's check whether the `credentials` dictionary has a key called `"user"`:

   ```
   if "user" in credentials:
   ```

3. Check whether the `token` key provided by the client matches the stored `token` key for the user:

```
if credentials['token'] == logged_users
    [credentials['user']]:
```

4. Create a variable called `token` to store the `token` key we found in the `logged_users` variable. Then, let's return the user's authentication `token` key by calling the `peer.put_var()` method and passing the JSON-formatted token string so that the client receives a response from the server:

```
var token = logged_users[credentials['user']]
peer.put_var(JSON.stringify(token))
```

Our function should look like this:

```
func get_authentication_token(peer, message):
    var credentials = message
    if "user" in credentials:
        if credentials['token'] == logged_users
            [credentials['user']]:
            Var token = logged_users[credentials
                ['user']]
            peer.put_var(JSON.stringify(token))
```

Now, whenever we need to make any procedure that requires confirmation from the server that the player is still in a valid play session, we can call this function. But to actually do that, we need to add two lines of code to our server so that it understands when the client makes such a request.

In the `_process()` function, we check whether the client is making a request for the `authenticate_credentials()` method. Let's check whether the client is making a request for the `get_session_token()` method instead, and if so, we call it. The `_process()` function should look like this:

```
func _process(delta):
    server.poll()
    if server.is_connection_available():
        var peer = server.take_connection()
        var message = JSON.parse_string(peer.get_var())
        if "authenticate_credentials" in message:
            authenticate_player(peer, message)
        elif "get_authentication_token" in message:
            get_authentication_token(peer, message)
```

Now, let's move on to the final part of our little project, where we are going to provide and load the player's avatar data.

Loading the player's avatar

Welcome to **AvatarScreen**! This is where the player will be able to customize their avatar appearance and select a unique name in the final version of our (fake) game. To display their current available avatar, we need to load the player's avatar data from the database and display it on the screen.

For that, the **AvatarScreen** scene is made up of a `Control` node called **AvatarScreen**, which holds all the other nodes in the scene, including a `Control` node called **AvatarCard**:

Figure 2.5 – The AvatarScreen scene's node hierarchy

The **AvatarCard** node contains a **TextureRect** node to display the avatar's image using a texture file and a **Label** node to display the avatar's name.

To load the player's avatar, we first need to retrieve the path to the image file from our fake database, which we previously populated with avatar information. So, before we dive into the action in the **AvatarScreen** logic, let's create the avatar retrieving logic in the `Server.gd` script, and let's work on the `get_avatar()` function this time. Proceed as follows:

1. Inside the `get_avatar()` function, create a local `dictionary` variable that contains the contents of the message:

```
func get_avatar(peer, message):
    var dictionary = message
```

2. Check whether there's a `"user"` key present in the `dictionary` variable:

```
    if "user" in dictionary:
```

3. If we find the `"user"` key in this dictionary, let's create a local `user` variable that is equal to the value of the user key in the `dictionary` variable:

```
        var user = dictionary['user']
```

4. Check whether the `'token'` key in the `dictionary` variable matches the token stored in the `logged_users` dictionary for the user specified by the `user` key:

```
if dictionary['token'] == logged_users[user]:
```

5. If this is the case, create a local `avatar` variable that is equal to the value of the `'avatar'` key in the `fake_database` dictionary for the user specified by the `'user'` key:

```
var avatar = fake_database[dictionary
    ['user']]['avatar']
```

6. Create a local `nick_name` variable that is equal to the value of the name key in the `fake_database` dictionary for the user specified by the `user` key:

```
var nick_name = fake_database[dictionary
    ['user']]['name']
```

7. Create a `response` dictionary with the `avatar` and `name` keys and values of `avatar` and `nick_name` respectively:

```
var response = {"avatar": avatar, "name":
    nick_name}
```

8. Use the `peer.put_var()` method to send the `response` dictionary as a JSON string to the client:

```
peer.put_var(JSON.stringify(response))
```

With that, we wrapped up our server, so we are ready to move to the **AvatarScreen** node. But before we do that, let's see how the `get_avatar()` function looks after our work:

```
func get_avatar(peer, message):
    var dictionary = message
    if "user" in dictionary:
        var user = dictionary['user']
        if dictionary['token'] == logged_users[user]:
            var avatar = database[dictionary
                ['user']]['avatar']
            var nick_name = database[dictionary
                ['user']]['name']
            var response = {"avatar": avatar, "name":
                nick_name}
            peer.put_var(JSON.stringify(response))
```

Now, let's open the `AvatarScreen.gd` script so that we can finally display our player's avatar! Go ahead to the `request_authentication()` function because, as mentioned before, every time we need to perform operations on the player's data, we need to verify their credentials.

9. Inside the `request_authentication()` function, create a variable called `request` that holds a dictionary with the `'get_authentication_token'`, `'user'`, and `'token'` keys. The value of `'get_authentication_token'` should be set to `true` just so that the server understands the request, while the values of `"user"` and `"token"` should be retrieved from the `AuthenticationCredentials` singleton:

```
func request_authentication(packet):
    var request = {'get_authentication_token': true,
        "user": AuthenticationCredentials.user, "token
        ": AuthenticationCredentials.session_token}
```

10. Use `packet` to send this request to the server by encoding the request as a JSON string using `JSON.stringify()` and then using the `put_var()` method to send it:

```
packet.put_var(JSON.stringify(request))
```

11. Use a `while` loop to wait for a response from the server. Inside the loop, create a variable called `data` to store the JSON response from the server, decoded using `JSON.parse_string()`:

```
while packet.wait() == OK:
    var data = JSON.parse_string(packet.get_var())
```

12. Check whether the `data` variable is equal to the `session_token` variable stored in the `AuthenticationCredentials` singleton. If it is, call the `request_avatar` function and break out of the loop:

```
if data == AuthenticationCredentials.
    session_token:
    request_avatar(packet)
    break
```

At the end, our `request_authentication()` function should look like this:

```
func request_authentication(packet):
    var request = {'get_authentication_token': true,
        "user": AuthenticationCredentials.user,
            "token": AuthenticationCredentials.
                session_token}
    packet.put_var(JSON.stringify(request))

    while packet.wait() == OK:
        var data = JSON.parse_string(packet.get_var())
        if data == AuthenticationCredentials.
            session_token:
            request_avatar(packet)
            break
```

It's finally time to retrieve the player's avatar data and display their avatar so that they can engage in our game world! For that, let's go to the request_avatar() function and create the avatar request and creation displaying logic.

13. Inside the request_avatar() function, create a dictionary named request with the 'get_avatar', 'token', and "user" keys and their respective values. We get the user and session tokens from the AuthenticationCredentials Autoload:

```
func request_avatar(packet):
    var request = {'get_avatar': true, 'token':
        AuthenticationCredentials.session_token,
            "user": AuthenticationCredentials.user}
```

14. Use the packet.put_var() method to send the request dictionary as a JSON-formatted string to the server:

```
packet.put_var(JSON.stringify(request))
```

15. Create a while loop to wait for the server to respond. Inside the loop, parse the response as a dictionary using the JSON.parse_string method and store it in a variable named data:

```
while packet.wait() == OK:
    var data = JSON.parse_string(packet.get_var())
```

16. Check whether the dictionary data contains the "avatar" key. If it does, load the texture of the avatar image from the path in the "avatar" key value, and set it as the texture of the **TextureRect** node, called texture_rect. Also, set the value of the **Label** node called label to the value of the "name" key in the data dictionary. Finally, exit the while loop with break:

```
if "avatar" in data:
    var texture = load(data['avatar'])
    texture_rect.texture = texture
    label.text = data['name']
    break
```

We have almost finished our login screen! Before we add the final touch, let's see how the request_avatar() method ended up:

```
func request_avatar(packet):
    var request = {'get_avatar': true, 'token':
        AuthenticationCredentials.session_token,
            "user": AuthenticationCredentials.user}
    packet.put_var(JSON.stringify(request))

    while packet.wait() == OK:
        var data = JSON.parse_string(packet.get_var())
```

```
        if "avatar" in data:
            var texture = load(data['avatar'])
            texture_rect.texture = texture
            label.text = data['name']
            break
```

17. Now, the final touch is to add yet another check on the `Server.gd` script to handle when we receive an avatar request. So, the `_process()` method should become something like this:

```
func _process(delta):
    server.poll()
    if server.is_connection_available():
        var peer = server.take_connection()
        var message = JSON.parse_string
            (peer.get_var())
        if "authenticate_credentials" in message:
            authenticate_player(peer, message)
        elif "get_authentication_token" in message:
            get_authentication_token(peer, message)
        elif "get_avatar" in message:
            get_avatar(peer, message)
```

And if we test our game by hitting the **Play** button, or if we test the **MainMenu** scene, we can verify that our game is working!

18. The first thing we need to do is to select the **Server** button in one of the debugging instances:

Figure 2.6 – Pressing the Server button in the MainMenu scene

19. Then, in another instance, choose **Client**, and it should immediately open the **LoginScreen** scene:

Figure 2.7 – Inserting the player's username into the client's LoginScreen UserLineEdit

20. Choose one of the users we have available in our fake database and insert their credentials:

Figure 2.8 – Inserting the player's username into the client's LoginScreen PasswordLineEdit

21. As soon as you press the **Login** button with the correct credentials, it should load the **AvatarScreen** scene with the respective avatar:

Figure 2.9 – The AvatarScreen scene displaying the player's avatar after a successful authentication

Congratulations! You've made your first login screen with authentication features, serializing and deserializing players' data all across the network. Be proud of yourself—this is a great feat!

Summary

In this chapter, we saw how we can establish a connection between server and client using the UDP protocol implementation in Godot Engine's network API. With that, the network peers can open a communication channel and exchange data.

Since this implementation works on quite a low-level approach, we saw how we can create a simple API for our peers to make, understand, and reply to each other's requests. Depending on the request, it might be necessary to follow a process known as serialization, which is how we take relevant information from our game state and turn it into a format that we can store and pass around. In our case, we saw that JSON format is one of the most common serialization formats.

Using the JSON format, we saw how we can parse our Godot Engine string as JSON and also how to turn a JSON file into a dictionary that we can work with more efficiently using GDScript.

At the end of the chapter, we saw how we can authenticate players' credentials, matching them against a fake database. With successful authentication, we gathered players' data to display their respective avatars based on their data in our database.

In the next chapter, we are going to add a new complexity level by allowing multiple clients to log in to the same server and finally have a shared experience. For that, we will create a Lobby node that displays all logged players' names and avatars!

3

Making a Lobby to Gather Players Together

In the previous chapter, we discussed how to use UDP packets to exchange data between multiple players in a game. While this approach is highly efficient, it requires a lot of manual work to ensure that data is sent and received correctly. In this chapter, we will explore the high-level network API of the Godot Engine, which simplifies the networking process by providing a set of built-in tools and functions that can handle common network tasks.

Specifically, we will focus on the **ENetMultiplayerPeer API**, which is the Godot Engine's wrapper class for its ENet library implementation, and **Remote Procedure Call (RPC)**, a communication protocol that allows us to make calls to functions and methods on remote computers as if we were making them locally. We will use these tools to create a lobby, authenticate players, retrieve player avatar data from a fake JSON database, and synchronize all players when a player enters the lobby. We will explore the benefits of using RPCs instead of exchanging UDP packets, and how this approach can simplify the process of synchronizing game states between multiple players.

We will cover the following topics in this chapter:

- Calling functions remotely with RPCs
- Understanding the multiplayer authority
- Comparing UDP and ENet approaches
- Remaking the login screen with RPCs
- Adding the player's avatar
- Retrieving players' avatars
- Testing the lobby

By the end of this chapter, you will have a solid understanding of how to use the Godot Engine's high-level network API and RPCs to create a robust multiplayer lobby for your game.

Technical requirements

In this chapter, we will build yet another project using the Godot Engine. Remember, throughout this book, we are using Godot Engine version 4.0, so this is also a requirement.

This time around, we are going to use the files provided in the `res://03.making-lobby-to-gather-players` folder. So, if you don't have the project repository yet, download it through this link:

`https://github.com/PacktPublishing/The-Essential-Guide-to-Creating-Multiplayer-Games-with-Godot-4.0`

Then, with the project added to your Godot Engine's project manager, open the project and proceed to the `res://03.making-lobby-to-gather-players` folder.

Calling functions remotely with RPCs

In a network context, **RPC** stands for **remote procedure call**, which is a protocol that allows one program to call a function or procedure on another program running on a different machine or over a network. In the context of the Godot Engine, RPCs allow objects to exchange data between each other over the network, which is a key feature in creating multiplayer games.

To use RPCs in the Godot Engine, we need to use the `ENetMultiplayerPeer` API, which provides a high-level network interface for handling network connections and sending and receiving data, as well as managing RPCs. By using `ENetMultiplayerPeer`, we can easily send and receive RPCs and handle network communication more straightforwardly.

When exchanging data with RPCs, objects can exchange data through functions, which makes the process more straightforward compared to exchanging data using UDP packets. With UDP packets, we need to send packets requesting procedures and wait for a response, and only then can we get the data. This process can be complex and difficult to manage, as we saw in the previous chapter, especially in large games with many objects exchanging data.

One limitation of RPCs is that they don't allow the transmission of objects, such as nodes or resources, through the network. This can be a challenge in games that require the exchange of complex objects between different machines. However, there are workarounds to this limitation, such as sending serialized data or using custom serialization methods. We learned how to do that in *Chapter 2, Sending and Receiving Data*, so this won't be a problem to us.

RPCs are a powerful tool for creating multiplayer games, and using the `ENetMultiplayerPeer` API in the Godot Engine makes it easy to use them. Although there are limitations to RPCs, such as the inability to transmit objects over the network, they are still a crucial part of creating a robust and seamless multiplayer experience.

Introducing the @rpc annotation

Godot Engine 4.0 introduces a new feature called @rpc function annotation. An **annotation** is a special marker that we can add to a function in our code that provides additional information to the compiler or the Godot Engine. In the case of @rpc, this annotation is used to mark functions that can be called remotely over the network in a multiplayer game.

There are several options that we can add to the @rpc annotation, which control how the function is called and executed over the network.

Let's take a closer look at each option:

- We have the calling options, meaning what should happen when we make an RPC to this function:

  - call_remote: This option indicates that the function should only be called remotely on instances of the node in other peers' machines but not on the node's local instance.

  - call_local: This option indicates that the function should also be called locally on the current peer's instance of the node as well. This is useful when we need to sync all peers in the network, including the caller.

- Then, we have the caller option, meaning who can call this function remotely:

  - authority: This option indicates that the function should only be called by the multiplayer authority. We'll see more about that soon.

  - any_peer: This option indicates that the function can be called by any peer in the network. This is useful for functions that can be executed by multiple peers in a multiplayer game and not only the multiplayer authority.

- We also have options regarding the reliability of the remote data exchange when we make an RPC to this function:

  - reliable: This option indicates that the function should be executed reliably over the network, which means that the function call will be guaranteed to arrive at its destination. This is useful for functions that need to be executed without the risk of losing data.

  - unreliable: This option indicates that the function should be executed with low reliability, which means that there is a chance that some data may be lost or delayed over the network. This is useful for functions that can tolerate some loss of data.

  - unreliable_ordered: This option is similar to unreliable but ensures that function calls are executed in order over the network. This is useful for functions that need to be executed in a specific order but can tolerate some data loss.

We can also specify which connection channel the RPC should use to transmit its data. This is useful to prevent bottlenecking or to dedicate some channels for specific functions. For instance, we can pick a channel that uses `reliable` data, such as the transmission of a message between players. Then, we can have another channel that uses `unreliable_ordered` data, such as updating peers about their avatars' current position.

In that sense, we just need the most recent position to arrive; every other call with previous positions is irrelevant. So, while one channel waits for messages to arrive, the other is constantly receiving new updates about avatars' positions and neither of them blocks the other.

The order in which we pass these options doesn't matter for the Godot Engine itself. The only exception is the channel, which should always be the *last option* passed. Take the following example:

```
@rpc("any_peer", "unreliable_ordered")
func update_pos():
    pass


@rpc("unreliable_order", "any_peer")
func update_pos():
    pass
```

These two ways of establishing the RPC options are the same and both will work. Now, the following annotation has an issue:

```
# This wouldn't work
@rpc(2, "any_peer", "unreliable_ordered")
func update_pos():
    pass


# This would work
@rpc("any_peer", "unreliable_ordered", 2)
func update_pos():
    pass
```

Only the second RPC annotation will work, because we are passing the channel as the last argument of the annotation. In the first RPC annotation example, we are passing the channel as the first argument, so it won't work.

Now that we understand how we can mark a function as an `rpc` function and what options we can use to fine-tune it and achieve what we need in our game, let's see what we need to be able to call such functions and propagate them across the network.

What's necessary for an RPC?

To make RPCs in the Godot Engine, first, we must set up an ENetMultiplayerPeer connection, which manages the network connection and handles the transmission of data between nodes. We did that in the first chapter, but we will also go through the process here as well.

Once the ENetMultiplayerPeer connection is set up, we must ensure that the NodePath to the nodes receiving the RPC is exact. This means that the NodePath to the target node must be identical on all peers in the network. If the NodePath is not exact, the RPC may not be sent to the correct node or may not be sent at all.

We can establish a default name for our root nodes to avoid issues. We chose the root node to facilitate the logic afterward as it is the first node in the hierarchy.

For instance, in our upcoming **Lobby** project, all three scenes share a common root name, Main:

Figure 3.1 – The ServerLobby, LoginScreen, and ClientLobby scenes with their root node named Main

It's also important to note that every node in the network should share all the methods marked with @rpc, even if they are not called or used by anyone. This is because the Godot Engine requires all nodes to have access to the same set of methods to function properly. This can be a minor inconvenience for developers as it may bloat some classes with unnecessary methods, but it is necessary for Godot's networking system to work effectively.

For instance, these are the methods found in each of the scenes that compose our upcoming **Lobby** project. Note that they all share the highlighted methods on top of their non-RPC methods:

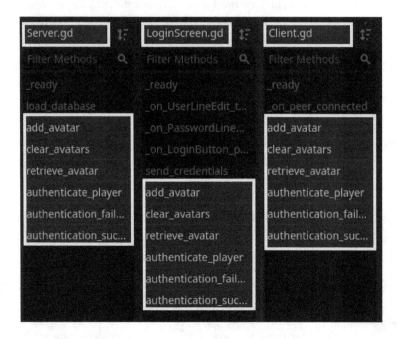

Figure 3.2 – The Server.gd, LoginScreen.gd, and Client.gd scripts
share functions marked with the @rpc annotation

Making RPCs in the Godot Engine requires setting up ENetMultiplayerPeer connections, ensuring that the NodePath to the target node is exact, and ensuring that all nodes in the network share all the methods marked as RPC. While this may require some additional setup and minor inconvenience, it enables developers to create multiplayer games easily and efficiently.

With that, we understand the core of RPCs in the Godot Engine. We saw what we need to set our game to support RPCs, how the @rpc annotation works, and how we can tweak it to match our design. In one of those tweaking options, we saw that it's possible to only allow the multiplayer authority to call a given RPC function. Let's see what the multiplayer authority is and what we can do with it.

Understanding the multiplayer authority

In Godot Engine's high-level network API, the **multiplayer authority** is a concept that refers to the node that has the authority to make decisions about a node state in a multiplayer game. When two or more peers are connected in a multiplayer game, it is important to have a centralized peer that decides what changes are valid and should be synchronized across all connected clients.

The multiplayer authority is assigned to a specific peer in the game, usually the server or host, and this peer has the power to decide which changes from a given node should be accepted and synchronized across all connected clients. This is important because in a multiplayer game, multiple players may try to make changes to the game state at the same time, and it is the responsibility of the multiplayer authority to manage, verify, and synchronize these changes correctly.

Each connected client in a multiplayer game is assigned a unique peer ID, which is a number that identifies the client within the game's network. Peer IDs are managed by the `Node.multiplayer.multiplayer_peer` object, which is a reference to the `ENetMultiplayerPeer` object, which handles the game's network connection. The `multiplayer_peer` object can be used to send and receive data between connected clients, as well as to manage the state of the game's network connection.

We can use the `Node.get_multiplayer_authority()` method to retrieve the node's current multiplayer authority, as well as set a different one using `Node.set_multiplayer_authority()`, passing the peer ID as an argument. Changing a node's multiplayer authority will allow the new peer to make and sync changes to this node across the network, and this can be quite dangerous.

For instance, if a player is responsible for a node that contains its health, the player may somehow hack it and be able to self-manage their health, ultimately becoming immortal.

Comparing UDP and ENet approaches

The `UDPServer` and `PacketPeerUDP` classes are lower-level networking tools that allow for the exchange of data through UDP packets. This approach requires more work from us, as we must manage the sending and receiving of packets ourselves. For example, to create a login system using `UDPServer` and `PacketPeerUDP`, we would need to create a packet that contains the user's login information, send it to the server, and then wait for a response.

In *Chapter 2, Sending and Receiving Data* project, we saw how to use `UDPServer` and `PacketPeerUDP` to pass data around. We saw that using these classes, we can serialize data and deserialize it on each end of the system, both client and server. Using this approach, we need to poll packets and wait for requests and responses to arrive. It does the trick, but you saw it can get a bit complicated.

One advantage of using the `UDPServer` and `PacketPeerUDP` classes is that they provide more control over the networking process, which can be useful for more complex games that require fine-tuned networking. However, this approach is also more error-prone, as we must handle the sending and receiving of packets ourselves, which can lead to issues such as packet loss or out-of-order packets.

On the other hand, using `ENetMultiplayerPeer` and RPCs provides a higher-level networking solution that simplifies the process of creating a login system. With this approach, developers can use the `@rpc` function annotation to mark a method as an RPC, which allows it to be called from any node in the network.

For example, to create a login system using `ENetMultiplayerPeer` and RPCs, we can mark the method that handles the login process as an RPC, and then call it from the client nodes. We are going to see that in a moment, and you will understand how powerful and simple the Godot Engine's high-level network API is.

Using `ENetMultiplayerPeer` and RPCs simplifies the networking process and makes it easier to create multiplayer games. The built-in features of `ENetMultiplayerPeer`, such as automatic packet ordering and error correction, make it easier to create a stable network connection. On top of that, the `@rpc` annotation makes it easy to call methods from any node in the network, simplifying the development process.

While the `UDPServer` and `PacketPeerUDP` classes provide more control over the networking process, using `ENetMultiplayerPeer` and RPCs offers a simpler and more streamlined approach to creating multiplayer games. The choice ultimately depends on the specific needs of the game you are making, but in most cases, using the higher-level tools provided by the Godot Engine will lead to a faster and more efficient development process.

Now that we understand how the Godot Engine's high-level network API solves a lot of issues through the `ENetMultiplayerPeer` class and how it compares to the UDP approach, with major advantages, such as its ability to easily allow the RPC features we need to make our game easier, let's remake the login screen we made in *Chapter 2*, *Sending and Receiving Data*, using these new tools. This will allow us to use the high-level API while understanding the low-level approach and the advantages of using the high-level approach instead.

Remaking the login screen with RPCs

Welcome back to our studio, fellow network engineer! In *Chapter 2*, *Sending and Receiving Data*, we learned how to create a basic login system using the Godot Engine's `UDPServer` and `PacketPeerUDP` classes. While this approach was perfect for our small-scale project, we need to level our game as we move forward and create a lobby!

Fear not, for we have the perfect solution for you – the Godot Engine's `ENetMultiplayerPeer` and RPCs! These two powerful tools will help us build a robust and efficient system that can easily scale up to support multiple connected clients – as far as we researched, up to 4,095 simultaneously connected players!

With the Godot Engine's `ENetMultiplayerPeer`, we can easily manage multiple connections and synchronize game data across all connected clients. This means that our login system will be able to handle more connections, and our game will run smoother than ever before!

With that, we will also be able to make RPCs! RPCs are an essential part of networking in the Godot Engine. They allow us to call functions on other nodes in the network as if they were local functions. With RPCs, we can easily share data and perform actions across all connected clients, making our login system even more robust and efficient.

So, get ready to level up our game, network engineer! In the upcoming sections, we'll dive into implementing the new login system using `ENetMultiplayerPeer` and RPCs and synchronizing players' avatars into a lobby.

We'll also cover some best practices and tips for working with `ENetMultiplayerPeer` and RPCs to ensure our multiplayer game runs smoothly and efficiently. With these powerful tools at our disposal, we'll be able to create a multiplayer game that will wow players and leave them wanting more.

Let's start by establishing connections between our players and the server using the `ENetMultiplayerPeer` API.

Establishing an ENetMultiplayerPeer connection

Let's recap from the first chapter how to establish a connection using the high-level `ENetMultiplayerPeer` class. We'll start with the server.

This time, we will also add elements such as the fake database and the logged users from our project from *Chapter 2*, *Sending and Receiving Data*. This will allow us to authenticate players and keep track of who's connected and their session tokens. Well, without further ado, let's dive into it!

We'll start by setting up an ENet multiplayer server on port 9999, loading our fake JSON database file, and assigning the `peer` instance to the `multiplayer_peer` property of the node's `multiplayer` object so we can make RPCs. Remember, we can only carry out RPCs within an established ENet connection:

1. Open the `res://03.making-lobby-to-gather-players/LobbyServer.tscn` scene, and then open the **Main** node's script.

2. Declare a constant variable, `PORT`, and assign our default value of 9999. This variable will be used later to specify the port number on which the server will listen for incoming connections:

    ```
    const PORT = 9999
    ```

3. Use the `@export` decorator to create a new variable, `database_file_path`, which can be edited from the **Inspector** panel. This variable will store the path to the JSON file that contains our fake user database. We are using the same file from the previous chapter:

    ```
    @export var database_file_path = "res://02.sending-and-
    receiving-data/FakeDatabase.json"
    ```

4. Create a new `ENetMultiplayerPeer` instance and assign it to the `peer` variable. This will be our high-level network interface for sending and receiving data between clients and the server and making RPCs:

    ```
    var peer = ENetMultiplayerPeer.new()
    ```

5. Create an empty dictionary called `database` and an empty dictionary called `logged_users`. These variables will be used to store our fake user data and keep track of which users are currently logged in, respectively:

```
var database = {}
var logged_users = {}
```

6. In the `_ready()` callback, call `peer.create_server(PORT)` to create a new multiplayer server that listens for incoming connections on the port number specified by the PORT variable:

```
func _ready():
    peer.create_server(PORT)
```

7. Still in the `_ready()` callback, assign the `peer` to `multiplayer.multiplayer_peer`. This variable makes our `peer` object the default network interface for all nodes in the game:

```
func _ready():
    peer.create_server(PORT)
    multiplayer.multiplayer_peer = peer
```

8. Finally, still in the `_ready()` callback, make a call to the `load_database()` method. We are going to create this in a moment. We do that to have the database in memory from the start, as soon as the server is ready:

```
func _ready():
    peer.create_server(PORT)
    multiplayer.multiplayer_peer = peer

    load_database()
```

9. Now, define a new function, `load_database()`, that takes an optional argument, `path_to_database_file`. This function will be used to load the user data from the JSON file into our `database` dictionary:

```
func load_database(path_to_database_file =
    database_file_path):
```

10. Inside `load_database()`, open the file specified by `path_to_database_file` using `FileAccess.open()` and assign it to the `file` variable:

```
func load_database(path_to_database_file =
    database_file_path):
    var file = FileAccess.open(path_to_database_file,
        FileAccess.READ)
```

11. Get the contents of the file as text using `file.get_as_text()` and assign it to the `file_content` variable:

```
func load_database(path_to_database_file =
    database_file_path):
    var file = FileAccess.open(path_to_database_file,
        FileAccess.READ)
    var file_content = file.get_as_text()
```

12. Parse the contents of `file_content` as JSON using `JSON.parse_string()` and assign the resulting dictionary to `database`:

```
func load_database(path_to_database_file =
    database_file_path):
    var file = FileAccess.open(path_to_database_file,
        FileAccess.READ)
    var file_content = file.get_as_text()
    database = JSON.parse_string(file_content)
```

At this point, this is how our `LobbyServer.gd` should look like this:

```
extends Control
const PORT = 9999
@export var database_file_path = "res://
    02.sending-and-receiving-data/FakeDatabase.json"

var peer = ENetMultiplayerPeer.new()
var database = {}
var logged_users = {}

func _ready():
    peer.create_server(PORT)
    multiplayer.multiplayer_peer = peer

    load_database()

func load_database(path_to_database_file =
    database_file_path):
    var file = FileAccess.open(path_to_database_file,
        FileAccess.READ)
    var file_content = file.get_as_text()
    database = JSON.parse_string(file_content)
```

With that, it's time to get our hands dirty with the fun part and the core of this chapter. Up next, we are going to finally create the @rpc methods that we'll use across our upcoming classes.

Creating the RPC functions template

With that, we can start to define our @rpc methods so when we move on to LobbyLogin, we already know what we'll call and how it works. So, still in LobbyServer, let's create some RPC methods.

These methods are going to be used on LobbyLogin and LobbyClient as well. Remember, all classes that make RPCs should share the same RPC methods even if they don't use them.

So, let's create this interface:

1. The @rpc annotation on this line is an RPC annotation that tells Godot that this function is called remotely only by the multiplayer authority, which is the server itself. A remote call means that when LobbyServer makes an RPC to this function, it won't execute it on itself locally. We will use the add_avatar() method to add a new avatar to the game's lobby, and we will implement it on LobbyClient:

    ```
    @rpc
    func add_avatar(avatar_name, texture_path):
        pass
    ```

2. The clear_avatars() function will remove all avatars from the lobby. We use this function to clear all avatars from the game so we can sync with newer players. This is also a method that we'll implement on LobbyClient:

    ```
    @rpc
    func clear_avatars():
        pass
    ```

3. This @rpc("any_peer", "call_remote") annotation tells Godot any peer can remotely call this function. We'll use the retrieve_avatar() method to retrieve the texture path for a specific player's avatar. We'll implement this method in LobbyServer soon, and the LobbyClient is the one that's going to remotely call it:

    ```
    @rpc("any_peer", "call_remote")
    func retrieve_avatar(user, session_token):
        pass
    ```

4. The `authenticate_player()` method will authenticate a player using a username and password. We use this function to authenticate players' credentials and pair them with a session token on the `logged_users` dictionary. This is also a method from `LobbyServer`, but now it's `LobbyLogin` that's going to remotely call it:

```
@rpc("any_peer", "call_remote")
func authenticate_player(user, password):
    pass
```

5. Then, we use the `authentication_failed()` method to notify a player that their authentication failed. We will call this from `LobbyServer` on `LobbyClient` when the server can't authenticate the credentials the player sent.

Note that while every function marked with an `@rpc` annotation should be on all other classes it interacts with, those classes don't need to have the same options for their `@rpc`. You will understand this better when we jump into `LobbyLogin` and `LobbyClient`:

```
@rpc
func authentication_failed(error_message):
    pass
```

6. We also have `authentication_succeed()`. We call this function from `LobbsyServer` on the player's `LobbyClient` to tell them that their authentication succeeded, providing them with their session token:

```
@rpc
func authentication_succeed(user, session_token):
    pass
```

With that, we have all the RPC functions we are going to use in our lobby system. `LobbyServer`'s RPCs section should look like this:

```
@rpc
func add_avatar(avatar_name, texture_path):
    pass

@rpc
func clear_avatars():
    pass

@rpc("any_peer", "call_remote")
func retrieve_avatar(user, session_token):
    pass

@rpc("any_peer", "call_remote")
func authenticate_player(user, password):
    pass
```

```
@rpc
func authentication_failed(error_message):
    pass

@rpc
func authentication_succeed(user, session_token):
    pass
```

Our template is ready. It has the @rpc methods that the classes that comprise our lobby need to share to communicate in our network. Remember, this is a necessary step; even if some of the classes don't implement the method, they should at least share this interface. For instance, coming next, we are going to implement the authentication logic in the lobby server, but other classes only need the method signature for that to work. Let's see how this goes.

Authenticating the player

In this section, we will focus on authenticating the player in the lobby server. We will use the authenticate_player() RPC method that we previously defined in our server script to verify the player's identity and grant access to the lobby.

The authenticate_player() method will take a username and a password as arguments and will return either an error message or a session token. If the credentials are invalid, the method will make a remote call to the authentication_failed() method with an error message explaining the reason for the failure.

If the credentials are valid, the method will make a remote call to the authentication_succeed() method, passing a session token and returning it to the player's LobbyClient. The session token is a unique integer number that identifies the player and is used to authenticate the player in subsequent RPCs.

Let's see how we can implement this logic using the tools we have at our disposal in the Godot Engine:

1. Inside LobbyServer's authenticate_player() method, get the peer_id of the player who sent the authentication request using the multiplayer.get_remote_sender_id() method. This is how we identify who sent the request so we can properly respond to the request:

    ```
    func authenticate_player(user, password):
        var peer_id = multiplayer.get_remote_sender_id()
    ```

2. Check whether the user exists in the database dictionary. If they don't exist, call the authentication_failed RPC method on the peer_id with the message "User doesn't exist". For that, we can use the rpc_id() method, which makes an RPC directly to the peer with the given ID:

```
if not user in database:
    rpc_id(peer_id, "authentication_failed",
        "User doesn't exist")
```

3. If the user exists in the database, check whether the password matches the password associated with the user. If it does, generate a random token using the randi() built-in method:

```
elif database[user]['password'] == password:
    var token = randi()
```

4. Then, add the authenticated user to the logged_users dictionary and call the authentication_succeed RPC method on the peer_id passing the token as an argument:

```
logged_users[user] = token
rpc_id(peer_id, "authentication_succeed",
    token)
```

This is what this method should look like:

```
func authenticate_player(user, password):
    var peer_id = multiplayer.get_remote_sender_id()

    if not user in database:
        rpc_id(peer_id, "authentication_failed",
            "User doesn't exist")
    elif database[user]['password'] == password:
        var token = randi()
        logged_users[user] = token
        rpc_id(peer_id, "authentication_succeed", token)
```

Note how useful making RPCs is. We don't need to poll or wait for packets to arrive at the destination or be concerned with serializing function arguments. We don't even have to create a *request* API to detect what the requester is trying to achieve as we did previously. It's very straightforward, almost like making a local application where you have direct access to the objects.

Now, let's see how we call this function on LobbyLogin. I'll assume that you already understand how it connects to the server using the ENetMultiplayerPeer.create_client() method. If you have any doubts about that, refer to the first chapter; the procedure is the same.

LobbyLogin resembles the login from the previous chapter, so let's skip directly to the send_ credentials() method, where it communicates with LobbyServer. You'll notice it also has the RPC methods we saw in LobbyServer. In this case, they all have the default options since the server is the only one that should call these methods on it:

1. In the send_credentials() method, retrieve the text String property from the user_ line_edit node and store it in the user variable:

    ```
    func send_credentials():
        var user = user_line_edit.text
    ```

2. Then, do the same but with password_line_edit and store it in the password variable:

    ```
        var password = password_line_edit.text
    ```

3. Finally, make an RPC to the multiplayer authority calling the authenticate_player() method with the user and password arguments. This will make this call only on LobbyServer:

    ```
        rpc_id(get_multiplayer_authority(),
            "authenticate_player", user, password)
    ```

    This is what the LobbyLogin.send_credentials() method will look like in the end:

    ```
    func send_credentials():
        var user = user_line_edit.text
        var password = password_line_edit.text

        rpc_id(get_multiplayer_authority(),
            "authenticate_player", user, password)
    ```

Let's take a look at the authentication_failed() and authentication_succeed() methods just so we understand how they work and how we keep the players' authenticated credentials across the scenes.

authentication_succeed() takes one argument called session_token, which is passed by the server when it authenticates the player's credentials, as we saw previously.

Then, we update the AuthenticationCredentials.user and AuthenticationCredentials.session_token values using user_line_edit.text and the session_token argument. Just like in the previous chapter, AuthenticationCredentials is a singleton autoload that stores the player's username and session token so we can use it in further scenes.

Talking about further scenes, after updating the `AuthenticationCredentials` singleton, we change the scene to `lobby_screen_scene` using `get_tree().change_scene_to_file(lobby_screen_scene)`. That means the player has successfully logged in and we can take them to the game lobby:

```
@rpc
func authentication_succeed(session_token):
    AuthenticationCredentials.user = user_line_edit.text
    AuthenticationCredentials.session_token = session_token
    get_tree().change_scene_to_file(lobby_screen_scene)
```

As for `authentication_failed()`, we set `error_label.text` to the error message received from `LobbyServer`. This will display the error to the player:

```
@rpc
func authentication_failed(error_message):
    error_label.text = error_message
```

Now that we understand how both sides of this relationship communicate and what they do with the data they pass around and get from each other, it's time to move on and see how the game resolves this data and displays the players' avatars to each other, synchronizing new players across the network every time they join the session.

In the upcoming section, we will see what the lobby screen looks like on the player's end and how we load, display, and sync players' avatars across the network.

Adding the player's avatar

In any online game, the player's avatar is a crucial element that represents them in the virtual world. In the previous section, we successfully authenticated the player and saved their session token and username in our `AuthenticationCredentials` autoload. Now, it's time to use that information to display the player's avatar in the lobby.

To achieve this, we will retrieve the player's avatar information from our fake database and create a new `AvatarCard`, a custom scene with a `TextureRect` node to display the avatar's image and a label to show its name. This way, players will be able to easily identify each other and feel more connected to the game world.

For that, let's open the `LobbyClient.gd` script. Here, we are going to do three major things:

1. Retrieve the avatar information from the server by making an RPC to the `retrieve_avatar()` method.
2. Implement the `add_avatar()` method that `LobbyServer` calls after retrieving the avatar data.

3. Implement the `clear_avatars()` method that `LobbyServer` calls before adding a new avatar to the lobby.

We are going to start with the latter two, then we can move on to the `LobbyServer.gd` file again to implement the `retrieve_avatar()` method:

1. In the `add_avatar()` method, create a new instance of `avatar_card_scene`:

```
@rpc
func add_avatar(avatar_name, texture_path):
    var avatar_card = avatar_card_scene.instantiate()
```

2. Add the newly created instance of `avatar_card` to `avatar_card_container`. This is an `HBoxContainer` node inside a `ScrollContainer` node:

```
avatar_card_container.add_child(avatar_card)
```

3. Wait for the next frame to process before continuing the execution of the code. We do that because `AvatarCard` needs to be ready before we update its data:

```
await(get_tree().process_frame)
```

4. Call the `update_data()` method on the `avatar_card` instance to update its data using the arguments passed to the `add_avatar()` method. With that, the lobby will use `avatar_name` to display the player's avatar name and will load the image stored in the `texture_path` to display their avatar's image:

```
avatar_card.update_data(avatar_name, texture_path)
```

The whole `add_avatar()` method should look like this:

```
@rpc
func add_avatar(avatar_name, texture_path):
    var avatar_card = avatar_card_scene.instantiate()
    avatar_card_container.add_child(avatar_card)
    await(get_tree().process_frame)
    avatar_card.update_data(avatar_name, texture_path)
```

Using the `@rpc` annotation, we created a method that the game's server can call on clients to add a new player avatar to all players' lobby screens, but this causes a small issue. As it is, this method may add the avatars that were already in the lobby before the newer player joined it.

So, we need to first clear all previous avatars and then add all currently logged players' avatars again. This ensures the lobby has only the correct avatars.

In the coming section, we are going to create a method that will run through all current avatars and remove them to have an empty `HBoxContainer`that we can use to add new avatars.

Cleaning AvatarCards

As mentioned before, whenever the server adds a new avatar to the lobby, it first cleans the lobby and recreates all avatars from scratch. We are going to see that in detail when we implement the `retrieve_avatar()` method.

The `clear_avatars()` method frees all existing avatars from the `avatar_card_container` node. It iterates over all children of `avatar_card_container` and calls `queue_free()` on each of them. After this function is executed, all avatars previously displayed in the lobby are removed from the container.

In the `clear_avatars()` method, iterate over each child node in `avatar_card_container` using a `for` loop and call the `queue_free()` method on each child node to remove it from `SceneTree` and free its resources:

```
@rpc
func clear_avatars():
    for child in avatar_card_container.get_children():
        child.queue_free()
```

That's it; pretty simple, right?

Now, before we move back to `LobbyServer.gd`, let's make an RPC to the multiplayer authority so it retrieves the current player's avatar. We do that in the `_ready()` method:

```
func _ready():
    rpc_id(get_multiplayer_authority(), "retrieve_avatar",
        AuthenticationCredentials.user,
            AuthenticationCredentials.session_token)
```

We use the `rpc_id()` method to call the `retrieve_avatar()` RPC method on the multiplayer authority, which is `LobbyServer` in this case. We pass the player's `username` and `session_token`, which are stored in the `AuthenticationCredentials` singleton autoload, as arguments to the `retrieve_avatar()` method. Now, it's time to move back to `LobbyServer.gd`.

Retrieving players' avatars

In this section, we will implement the `retrieve_avatar()` method on the `LobbyServer.gd` script, which will allow players to request their avatar data from the server. The avatar data is stored in the fake database. The server will respond with some RPCs to update all players with the appropriate data, displaying their avatars in the shared lobby.

With this method in place, we will complete the functionality of our **Lobby** project. Players will be able to authenticate themselves and display their avatars in the lobby. This will provide a solid foundation for building more complex multiplayer games in the upcoming chapter, as the basics of networking have been covered.

Let's do it!

1. In the `retrieve_avatar()` method, check whether the user is logged in by verifying that the user exists in the `logged_users` dictionary. If the user is not logged in, exit the function:

    ```
    func retrieve_avatar(user, session_token):
        if not user in logged_users:
            return
    ```

2. Then, check whether the session token provided by the remote peer matches the session token stored in the `logged_users` dictionary for the user:

    ```
    if session_token == logged_users[user]:
    ```

3. If the tokens match, call the `clear_avatars()` function on all connected peers to clear any existing avatars from their lobby screen:

    ```
    rpc("clear_avatars")
    ```

4. Loop through all the logged-in users stored in the `logged_users` dictionary:

    ```
    for logged_user in logged_users:
    ```

5. Retrieve the avatar name and texture path for the current `logged_user` from the `database` dictionary:

    ```
    Var avatar_name = database[logged_user]
        ['name']
    var avatar_texture_path = database
        [logged_user]['avatar']
    ```

6. Call the `add_avatar()` method on all connected peers and pass in `avatar_name` and `avatar_texture_path` as arguments to display the avatar in the lobby:

    ```
    rpc("add_avatar", avatar_name,
        avatar_texture_path)
    ```

    This is how the `retrieve_avatar()` method should look after all these steps:

    ```
    @rpc("any_peer", "call_remote")
    func retrieve_avatar(user, session_token):
        if not user in logged_users:
            return
        if session_token == logged_users[user]:
    ```

```
rpc("clear_avatars")
for logged_user in logged_users:
    var avatar_name = database
        [logged_user]['name']
    Var avatar_texture_path = database
        [logged_user]['avatar']
    rpc("add_avatar", avatar_name,
        avatar_texture_path)
```

Pay attention to its @rpc annotation options. Notice that any peer can call it remotely. This is how we make an RPC API for our online multiplayer games in the Godot Engine.

Some methods should be called remotely only by the multiplayer authority, some should be called locally as well, and some can be called by any peer on the network. It is up to us to decide and manage how peers interact with each other.

With all that in place, it's time to test out our game using multiple instances to simulate a server and multiple players connected to our network. Let's do this in the next section!

Testing the lobby

To test this out, we are going to run three instances of the game:

1. Go to **Debug | Run Multiple Instances** and select **Run 3 Instances**.

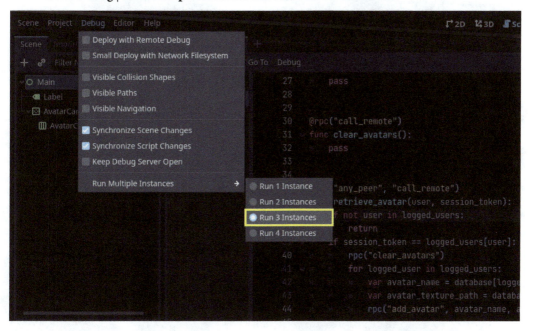

Figure 3.3 – Choosing to run three instances in the Run Multiple Instances menu

2. Then, open the res://03.making-lobby-to-gather-players/MainMenu.tscn scene and hit the **Play** button.

3. Pick one of the instances to be the game's server. To do that, just click on the **Server** button.

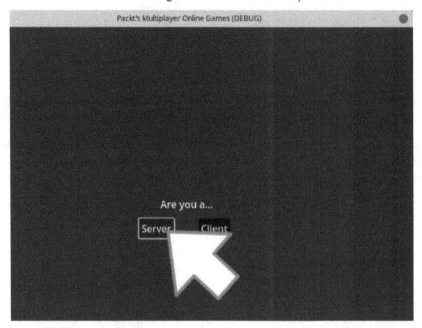

Figure 3.4 – Pressing the Server button on the MainMenu screen

4. Now, pick another instance and click on the **Client** button. It will take you to the LobbyLogin screen, where you can enter the first fake player's credentials.

5. Insert user1 in the username field and test in the password field. These are the credentials we added to FakeDatabase.json for our first user. Then, press the **Login** button. It will take you to the LobbyClient screen with a single avatar.

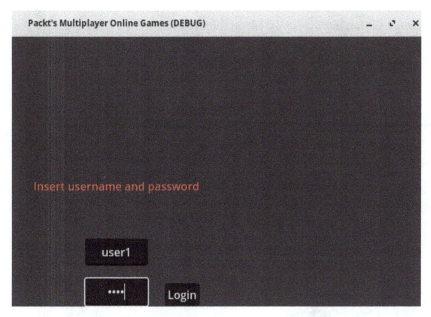

Figure 3.5 – The LobbyLogin screen with the player's credentials

With that, the server will authenticate the player's credentials and will allow the player to move on to the next screen, displaying the player's character's avatar and name based on the data it matched in the database file. In the following screenshot, we can see the next screen after a successful login.

Figure 3.6 – LobbyClient displaying the player's avatar after login

6. Then, select the last instance and click on the **Client** button as well. On the LobbyLogin screen, use the second player's credentials. In the first field, insert user2, and then, in the second field, test. It will take you to the LobbyClient screen, where there should be two avatars now. You can check the other client instance and they will both have the same avatars in order.

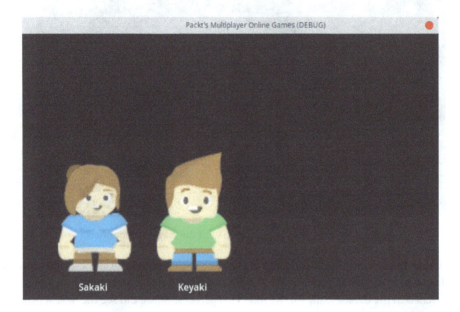

Figure 3.7 – The game displaying both players' avatars after the second player is logged in

We can see that everything is working as we want! The players can insert their credentials and the server authenticates them and provides a session token to keep them logged in after their validation. When logged in, they can see their avatar. Not only that, but our game also syncs players' avatars when a new player joins the session.

We did all that using the powerful @rpc annotation, which is possible to use when peers connect using the ENetMultiplayerPeer API.

Summary

In this chapter, we learned about RPCs and their importance in multiplayer game architectures. We saw how RPCs can be used to exchange data between nodes in the Godot Engine. We also saw what a multiplayer authority node is and how to set one up that manages all the game states between network peers. On top of that, we saw that by using the multiplayer API and ENetMultiplayerPeer, we can easily handle the communication between nodes.

Throughout the chapter, we created a lobby, which is a multiplayer game that features a lobby where players can join together. We saw how to create a client-server architecture, authenticate users, and exchange data between the server and the clients using RPCs. We also learned how to use the multiplayer API and `ENetMultiplayerPeer` to create a connection between the client and the server.

One of the essential concepts we learned is how `ENetMultiplayerPeer` simplifies the whole process of creating a multiplayer game compared to the low-level UDP approach. It abstracts away the complexity of low-level network programming, such as sending and receiving data packets, managing connections, and handling errors. This makes it easier for us to focus on implementing the gameplay mechanics of the game rather than worrying about the low-level details of the network communication.

Overall, this chapter has provided a solid foundation for developing multiplayer games in the Godot Engine. By following the steps outlined in this chapter, developers can create a simple lobby-based multiplayer game that utilizes RPCs, authentication, and the multiplayer API.

In the upcoming chapter, we are going to test the `ENetMultiplayerPeer` capabilities to exchange, update, and sync players. For that, we will create a chat room where players can communicate with each other and finally create a shared experience and a sense of community.

4

Creating an Online Chat

Welcome to the next chapter of our book on making online multiplayer games using Godot Engine 4.0!

In the previous chapter, we saw how to make a lobby for players and its importance to gathering before entering a game. Now, we will take a step further and explore the development of an online chat using Godot's Network API.

As we saw, Godot's Network API provides us with a robust set of tools for building real-time multiplayer games. In this chapter, we will use the `ENetMultiplayerPeer` class to establish a reliable connection between players, and some **remote procedure call** (**RPC**) methods to handle the chat system's logic.

As you may already know, a chat system is essential in any online multiplayer game, as it enables players to communicate with each other during gameplay. A well-designed chat system can greatly enhance the player experience, allowing for smoother coordination between team members and encouraging socialization among players.

In this chapter, we'll also explore the importance of using multiple communication channels in our network. We introduced this concept in the previous chapter, but now we are going to see how to actually use this feature in RPC methods so that our network is able to smoothly pass data around peers. By the end of the following step-by-step instructions, we will end up with the following:

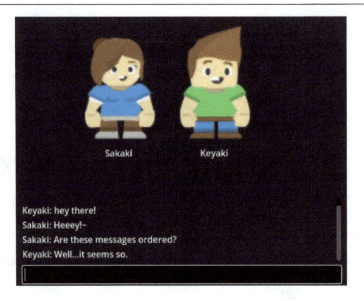

Figure 4.1 – The Chat screen with players exchanging messages

We will cover the following topics in this chapter:

- Understanding data exchange and channels
- Sending chat messages
- Updating peer's data remotely

Technical requirements

Through this chapter, we are going to use the fourth folder of our Godot Engine project repository, available through the following link:

`https://github.com/PacktPublishing/The-Essential-Guide-to-Creating-Multiplayer-Games-with-Godot-4.0`

After importing the project into your Godot Engine project manager, open it and navigate to the `res://04.creating-an-online-chat` folder. Then, open the `ChatControl.tscn` and `ChatControl.gd` files. They are going to be the focus of this chapter.

Coming next in this chapter, we are going to learn the basic concepts of reliable and unreliable data exchange and how channels work so we have the ground set for our chat system.

Understanding data exchange and channels

A chat system is essentially a chronological stack of messages that we order based on what players send to each other. Since this system needs a coherent chronological order, we need to understand how to prevent data from getting mixed and disorganized as it gets transmitted throughout the network. We also need to prevent this ordering from impacting other systems and components. To do that, we are going to learn how we can send packets reliably and how we can use multiple channels for data exchange.

Reliable and unreliable packets

Godot Engine's Network API allows for reliable and unreliable data exchange between peers. The `@rpc` annotation provides a way to transmit data securely between clients and the server using different transport protocols such as UDP and TCP. Reliable data exchange ensures that data is delivered in order and is not lost in transit, making it ideal for crucial data such as chat messages. Unreliable data exchange is faster and more efficient but offers no guarantee of order or delivery, making it ideal for non-critical data such as player position or real-time updates. In our chat system, we use the reliable `@rpc` option for chat messages to ensure that they are delivered in a timely manner without loss or duplication. Reliable data exchange ensures that players are able to follow the conversation and respond appropriately. Channels provide another layer of control over data exchange, allowing for better network optimization by prioritizing or separating data sent through different channels.

In the upcoming section, let's learn how channels work and what we can do with them.

Understanding channels

When developing multiplayer games in Godot Engine, understanding how communication channels work within the Network API is crucial for optimizing network performance and minimizing latency. In the context of Godot Engine's Network API, communication channels are used to separate different types of data being exchanged between peers. For example, you may want to use one channel for game state updates, another channel for chat messages, and another channel for player movement data.

The `@rpc` annotation in Godot Engine's Network API provides an option to specify the channel that an RPC method should use for sending and receiving data. By default, all RPC methods use channel 0. However, we can specify a different channel by passing an integer number as the last option in the `@rpc` annotation. For example, if you want to use a channel for game state updates, you can assign game state updates to channel 1 and chat messages to channel 2.

Using multiple channels in your Godot Engine multiplayer game can help improve network performance and minimize latency. By separating different types of data into separate channels, you can prioritize more important data and prevent congestion on a single channel.

It's worth noting that using multiple channels can also help prevent data loss and corruption in the event of packet loss or network congestion. By separating data into different channels, you can ensure that data loss or corruption on one channel doesn't affect the other channels. This can help prevent issues such as desynchronization between peers or corrupted game data. We are going to talk about that in the next section.

Remember, we are going to use reliable data exchange for our messages, so we can't prevent players from updating other crucial information, such as their and others' avatars' positions, just because the data exchange channel is still waiting for a chat message to arrive. It's wise to use another channel for that. Let's see how this works.

Open the `res://04.creating-an-online-chat/ChatControl.tscn` scene and you will notice it's already structured with all the necessary nodes we need to make our chat system. So, we are going to focus on the script here. The following figure shows the `ChatControl` scene node hierarchy:

Figure 4.2 – The ChatControl scene node hierarchy

From there, open the `res://04.creating-an-online-chat/ChatControl.gd` file, and make the method responsible for adding messages to the chat:

1. Create an `@rpc` annotation on top of the `add_message()` method. This RPC method should be available to any peer; it should also call itself locally. On top of that, it'll be a reliable method. Finally, here, we are going to use a separate channel to exchange data, so let's use channel 2:

    ```
    @rpc("any_peer", "call_local", "reliable", 2)
    func add_message(_avatar_name, message):
        pass
    ```

2. Inside the `add_message()` method, create a new variable called `message_text` that's going to use the `_avatar_name` String and the message arguments to create a text that uses two placeholders separated by a colon, like this:

    ```
    var message_text = "%s: %s" % [_avatar_name,
        message]
    ```

3. Then, concatenate the `label.text` String, skip a line, and then add the `message_text` String. This will add the latest message to the players' visible chat:

```
label.text = label.text + "\n" + message_text
```

4. Finally, to ensure the players' chat is always showing the latest message, we update `container.scroll_vertical` to match the `label.size.y float`. This way, it will scroll to the bottom of the chat label, displaying the latest message:

```
container.scroll_vertical = label.size.y
```

The complete `add_message()` method should look like this at this point:

```
@rpc("any_peer", "call_local", "reliable", 2)
func add_message(_avatar_name, message):
    var message_text = "%s: %s" % [_avatar_name,
        message]
    label.text = label.text + "\n" + message_text
    container.scroll_vertical = label.size.y
```

With that, we can use a dedicated channel to transmit our players' messages from one to another and display them in each player's chat interface. Using an independent data transmission channel is as simple as adding an integer number as the last option of a function's `@rpc` annotation.

In the next section, let's learn how we can gather the player's message and actually process it, sending it to other players throughout our peers' network.

Sending chat messages

Godot Engine's RPCs allow for efficient data transmission between clients and the server in multiplayer games. We can create an RPC method for this specific purpose, with message data as arguments. The transmission can be either reliable or unreliable, depending on the needs of the application. Once the message is sent, it's received by the appropriate recipients (including clients and the server) who handle it appropriately, such as by displaying it to the user or logging it. We did that in the *Understanding data exchange and channels* section when we made the `add_message()` method.

Sending messages using Godot's RPCs is a straightforward process that involves defining the message format. In our case, we use the player's avatar name and the message content, as seen previously, calling an RPC method to transmit the message and handling the message appropriately on the receiving end.

We are going to implement a method to read the message that players input using the `LineEdit` node and send it to all peers on the network. For that, we are going to use the `LineEdit.text_submitted` signal connected to the `_on_line_edit_text_submitted()` callback method, as shown in the following image:

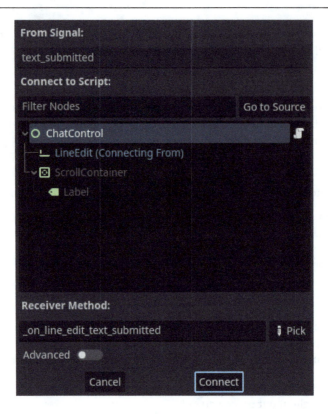

Figure 4.3 – The LineEdit text_submitted signal connection

With the signal connected, open the script and let's work on the _on_line_edit_text_
submitted() method:

1. The first thing we need to do is to prevent the processing of empty messages. For this, let's
    return from the function if the new_text argument is an empty string:

    ```
    func _on_line_edit_text_submitted(new_text):
        if new_text == "":
            return
    ```

2. Then, we can make an RPC to the add_message() method. This will call this method on
    all connected peers. We are going to pass the ChatControl.avatar_name String and the
    new_text String as arguments so the peers have the proper data to create their chat messages:

    ```
    rpc("add_message", avatar_name, new_text)
    ```

3. Finally, we clear the line_edit.text String to visually communicate that the game received
    the player's message and is processing it:

    ```
    line_edit.clear()
    ```

The complete `_on_line_edit_text_submitted()` method should look like the following:

```
func _on_line_edit_text_submitted(new_text):
    if new_text == "":
        return
    rpc("add_message", avatar_name, new_text)
    line_edit.clear()
```

With this method, any player in the game will be able to input a message and ask all peers, including themselves, to add a new message to their chat based on the `add_message()` logic we saw previously.

Now, we need to understand how we are going to update each player about other players' messages. In the next section, we will learn how to use RPCs on nodes other than the root node of a scene. With that, we will be able to make more concise scripts since we won't bloat our classes with methods they won't implement.

Updating peer's data remotely

Something really cool about Godot Engine's Network API is that we can abuse RPCs to pass data around. For instance, we've seen that we use the player's avatar name in our messages. But have you asked yourself how we retrieve this data in any of these steps?

You probably saw that there's an RPC method called `set_avatar_name()`, right? Since its `@rpc` annotation doesn't have any options, you can assume that it uses the default options. This is important to know because, as we saw previously, this means that it should be called remotely only by the **Multiplayer Authority** – in this case, the server.

Let's open `ChatServer.gd` to understand what's happening behind the scenes. In essence, most of it is pretty much the same as in the **Lobby** project, but you will notice something slightly different in the `retrieve_avatar()` RPC method. In *line 39*, we have the following instruction:

```
var peer_id = multiplayer.get_remote_sender_id()
```

We saw that this is a way to keep the sender of the latest RPC in memory so that we can refer back to it if necessary, and in this case, it will be necessary.

In *line 45*, we make a `rpc_id()` call to the `set_avatar_name()` method on the peer that just requested its avatar's data:

```
chat.rpc_id(peer_id, "set_avatar_name", database[user]
    ['name'])
```

Note that there's something else on top of that. We are calling this `rpc_id()` method from the `ChatControl` node, which is a child node of the `Main` root node.

This is the server's own `ChatControl` node. Since the server and client's scene tree have the same `NodePath` to their `ChatControl` node, we can make this `rpc_id()` call on the server's `ChatControl`node instead of making it from the `Main` node, and it will remotely call it on the player who requested their avatar's data:

Figure 4.4– The ChatServer and ChatClient's scene node hierarchies

This is a good and simple way to prevent bloating a single class with many RPC methods because, remember, if a caller has an RPC method, all peers should have this method on their equivalent nodes at the same `NodePath`.

It's also a very effective way to remotely update nodes with new data. RPCs are really impressive and useful tools to have at our disposal when making online multiplayer games with Godot Engine.

Summary

Throughout this chapter, we saw how we can use RPC methods to pass data around and perform actions on multiple peers of our network. We also understood the core difference between reliable and unreliable data exchange and saw some examples of situations of when to use each one. Due to this core difference in the way we can exchange data between the peers of our network, we also understood that one way may block the other, so we can use channels to prevent that one type of data from getting in the way of another type of data unrelated to that exchange.

By creating an online lobby where players can chat, we saw how to use the `@rpc` annotation with some of its available options, including the option to allow other peers to make remote calls instead of only the Multiplayer Authority.

In the next chapter, we will use the knowledge we've just acquired to build an actual real-time multiplayer experience. We'll create a multiplayer online quiz where players will compete to see who can pick the correct answer the quickest. See you there!

Part 2: Creating Online Multiplayer Mechanics

After understanding the tools we have available to create online multiplayer games, we put them into context by creating actual game prototypes. In this part, we learn how to turn single-player games into online multiplayer games, starting with a quiz game and ending with a prototype of an MMORPG game.

This part contains the following chapters:

- *Chapter 5, Making an Online Quiz Game*
- *Chapter 6, Building an Online Checkers Game*
- *Chapter 7, Developing an Online Pong Game*
- *Chapter 8, Designing an Online Co-Op Platformer*
- *Chapter 9, Creating an Online Adventure Prototype*

5

Making an Online Quiz Game

In this chapter, we will dive into the fascinating realm of creating an online quiz game using the powerful Network API offered by Godot Engine 4.0. We will explore how to leverage Godot Engine's Network API to create an engaging and interactive quiz game that can be played with friends or strangers online. We will cover the fundamental concepts of online multiplayer game development, including client-server architecture, game synchronization, and player interactions.

Here, we won't go through the game design aspects of this type of game: scoring, managing incentives, balancing, and so on; instead, we are going to focus on the engineering side of the equation: how to sync answers, prevent players from answering when another player has already answered, update question data for both players, and so on.

We will start by setting up the server side of the quiz game, including creating a dedicated server that can handle incoming connections and answers from multiple clients. We will then move on to designing the gameplay's core functionalities, including handling player input, and managing answers and quiz questions, while also handling communication between the clients and the server.

Throughout this chapter, we will learn how to use Godot Engine's **RPCs** to manage connections, handle data synchronization, and implement real-time multiplayer gameplay mechanics. We will also cover topics such as managing the game state.

We will cover the following topics in the chapter:

- Introducing the online quiz game
- Setting up a lobby for the quiz game
- Implementing online turns
- Turning local mechanics into remote gameplay

By the end of this chapter, you will have a solid understanding of how to create an online quiz game using Godot Engine 4.0's Network API. You will have learned the key concepts and techniques to make online asynchronous games, including server-side setup, client-side implementation, and network communication. The following screenshot showcases the end result of our online quiz game:

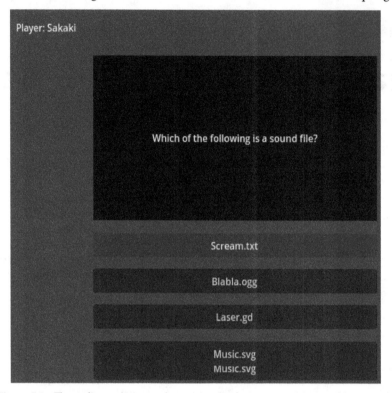

Figure 5.1 – The online quiz gameplay screen displaying a question and its options

In the next section, we are going to discuss the components of the quiz game so that we can pinpoint where we, as network engineers, have to implement the necessary features to turn the local multiplayer version of the game into an online multiplayer one.

Introducing the online quiz game

Welcome, network engineer! Our studio needs you to turn our quiz game into an online multiplayer experience! We have already gone through various challenges to create a captivating quiz game. Now, it's time to take it to the next level by adding online multiplayer functionality.

Imagine players from around the world competing against each other in real time, testing their knowledge and skills. In this chapter, you'll dive into the world of networking in game development and learn how to implement multiplayer features using GDScript. So, let's get started and make our quiz game an unforgettable multiplayer experience!

One of the key features of our online multiplayer quiz game is the dynamic loading of questions from a JSON database. Each question in the database contains all the necessary data, such as the question itself and four alternatives for players to choose from. Only one of the alternatives is the correct answer, and this information is stored in the database as well, ensuring fair and consistent gameplay.

To provide a simple and intuitive user interface, our game features four buttons, each representing an answer to the question displayed on the screen. A panel with a label displays game messages, including the current question that players must answer to score in the round. The game's interface is designed to provide a seamless experience for players as they navigate through the questions and answers.

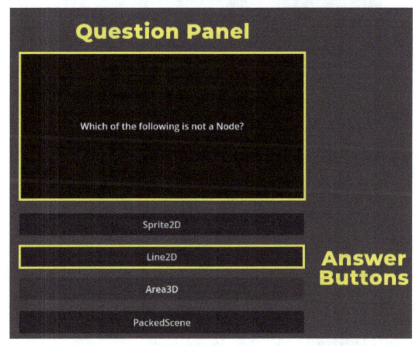

Figure 5.2 – The quiz screen displaying the round's question and available answers

As players correctly answer questions, they progress through the rounds. When a player wins a round, the game updates the question and answer options, ensuring that players are constantly challenged with new questions. The game continues until there are no more questions left to answer, making for an engaging and competitive multiplayer experience.

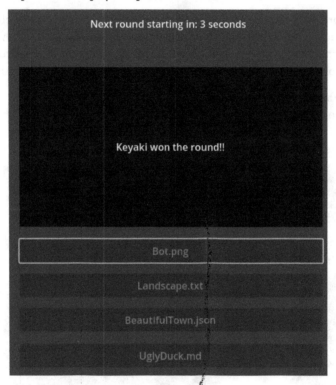

Figure 5.3 – The quiz screen displaying the round winner

In the next section, we will delve into the lobby of our quiz game. We will explore how to create a seamless multiplayer lobby system, allowing players to join games and compete in a fun and engaging multiplayer environment.

Setting up a lobby for the quiz game

In the upcoming section, we will be delving into the process of setting up a lobby for our quiz game. The QuizLobby scene serves as the hub for players to interact and prepare for the game.

Figure 5.4 – The login screen displaying the username and password fields and the players in the match

The process of authenticating players is similar to what we have done in previous chapters, utilizing data submitted by players and matching it against FakeDatabase. This ensures that only registered players with valid credentials are allowed to access the lobby.

Once a player successfully logs in, their name will appear for other players, providing visibility of the players currently present in the lobby. You can optionally add the previous chat to this scene as well to allow players to interact before the match starts. This will create a sense of community and allow players to connect and interact with each other while waiting for the game to start.

The QuizLobby scene is similar to our previous lobbies. So, in this section, we are going to focus on the core features it adds on top of the lobby we created in *Chapter 3, Making a Lobby to Gather Players Together*.

In the following section, we will create and understand the additional features in this new lobby iteration. For that, open the script at res://05.quiz-online/QuizLobby.gd and move on to the add_logged_player() method.

Displaying new players

One of the new features we will have in this updated lobby is the ability to see all the players who joined the current match. To implement that, let's take the following steps:

1. Inside the add_logged_player() method, set the logged_players_label.text property to player_name; this function receives an argument. The resulting text should append the player_name below the current content. For that, we concatenate the string with a placeholder string that skips a line and formats the placeholder as player_name:

    ```
    @rpc
    func add_logged_player(player_name):
        logged_players_label.text = logged_players_
            label.text + "\n%s" % player_name
    ```

2. After that, move on to the start_game() method and add the "authority" and "call_local" options to the @rpc annotation:

    ```
    @rpc("authority", "call_local")
    func start_game():
        pass
    ```

3. Then, inside the function, let's tell the SceneTree to change the current scene to quiz_screen_ scene, which is a variable that points to QuizScreenClient.tscn:

    ```
    @rpc("authority", "call_local")
    func start_game():
        get_tree().change_scene_to_file(quiz_screen_scene)
    ```

4. Finally, on the _on_StartButton_pressed() callback, we will make a direct RPC to the multiplayer authority's start_game() method without calling it locally:

    ```
    func _on_StartButton_pressed():
        rpc_id(get_multiplayer_authority(), "start_game")
    ```

In game development, ensuring fair gameplay and providing an enjoyable experience for players are critical aspects of creating a successful game. This involves implementing various features and functionalities that make the game engaging and dynamic. One such feature is adding a player to the match when they join the game. This can be achieved by creating a panel that displays the names of all the players currently participating in the game.

In a multiplayer game, communication between the server and clients is crucial. It's essential to ensure that only authorized entities can perform specific actions. This ensures that the game's mechanics and flow are consistent and reliable for all players.

Finally, once all the players are in the game, and the match is about to begin, the next step is to move them and the server to the next game screen. This screen will display all the necessary information about the game, such as the objective, rules, and game mechanics.

This ensures that all the players are on the same page and know what to expect from the game. Overall, implementing these features ensures that the game runs smoothly and that the players have a positive gaming experience.

With that, every time a player joins the match, their name will be added to the **Players in Match** panel. Note that regarding the `start_game()` method, only the multiplayer authority can make a remote call to it, right? We are about to see something new in a moment.

On the server side, we will have a different implementation of this method. In the upcoming section, we will see how the match actually starts and why we move all players and the server to the next game screen.

Starting the match

This implementation works that way to prevent one player from calling the `start_game()` method on the other players, or on itself, without other players also starting the game. The idea is that the player who presses the **Start** button will ask the multiplayer authority to start the game.

In turn, the multiplayer authority, which is the server in this case, will tell every player to also start the match. It does that by calling the `start_game()` method on each player. Let's see how this is done:

1. Open the script at `res://05.online-quiz/QuizServer.gd` and find the `start_game()` method.

2. In the `@rpc` annotation line, add the `"any_peer"` and the `"call_remote"` options. This will allow any peer on the network to make a remote call to this method:

    ```
    @rpc("any_peer", "call_remote")
    func start_game():
    ```

3. Then, tell the SceneTree to change to `quiz_screen_scene_path` using the `get_tree().change_scene_to_file()` method. This will tell the server to also update its context to the one in the `QuizScreenServer` scene. This will be necessary for the actual game to run:

    ```
    @rpc("any_peer", "call_remote")
    func start_game():
    get_tree().change_scene_to_file(quiz_screen_scene_path)
    ```

4.   Finally, and most importantly, make an RPC call to other peers' `start_game` method, so everyone in the network moves on to their respective `QuizScreenClient` scene:

```
@rpc("any_peer", "call_remote")
func start_game():
    get_tree().change_scene_to_file
        (quiz_screen_scene_path)
    rpc("start_game")
```

The lobby system is a vital component of any online multiplayer game, as it serves as the gateway for players to connect and prepare for the match. In our quiz game, we have successfully implemented the lobby system using Godot's built-in **Remote Procedure Call** (**RPC**) functionality. This feature allows us to establish a reliable two-way communication channel between the client and server, ensuring that all players are in sync.

With the lobby system in place, players can join the match and their names will be added to the **Players in Match** panel. It is important to note that the `start_game()` method can only be called by the multiplayer authority, preventing unauthorized calls and ensuring the integrity of the game. The server-side implementation of this method will be different, and we will explore this in the upcoming section.

If you want to add more features to the lobby system, you can create a countdown timer similar to the ones found in games such as *Warcraft III: Reign of Chaos*. This feature adds excitement and anticipation to the match and can help players prepare mentally for the upcoming game. However, for our quiz game, we are ready to move on to the next step.

With that, we have the lobby part of our quiz game ready to gather some players together and set them up, ready to start the match. We saw how to use the options that Godot provides for the `@rpc` annotation to create a two-sided communication that we can use to sync players and move them all together to the actual game.

The lobby system is a crucial part of any online multiplayer game, and we have successfully implemented it in our quiz game using Godot's built-in RPC functionality. The system allows players to join the match and sync their data with the server, ensuring that the game is fair and consistent. While we can add more features to the lobby system, such as a countdown timer, we are now ready to move on to the next stage of development.

In the next section, we are going to create a mechanism to disable players' actions when another player has already selected the correct answer. With that, you can even create a *turn-based* mechanism if you want, which is what we are going to do in *Chapter 6, Building an Online Checkers Game*.

Implementing online turns

When designing a quiz game, it's important to ensure that players can only provide one answer to a given question. This can be especially challenging when creating a multiplayer game, as multiple players may attempt to answer the question at the same time.

To prevent this, it's necessary to implement a system that disables players' ability to answer the question once a valid answer has been provided by another player.

One common approach to implementing this system is to disable the buttons representing the potential answers once a player has provided a response. This can be accomplished using code that identifies which button was pressed and compares it to the correct answer stored in the game's database. Once an answer has been identified, the code can disable the buttons and prevent other players from answering the question.

To further improve the player experience, it's also common to include a brief pause after an answer has been provided. During this time, players can review the questions and answers, and the game can display feedback on whether the answer was correct or not. This can help to build tension and excitement in the game, while also giving players a chance to reflect on their performance and improve their skills.

We need to prevent players' from answering the same question after another player already answered it. To do that, we can disable the buttons that represent answers once a valid response is provided. And for a better experience, we can add a brief pause before changing to the next question.

In this section, we are going to understand how we can prevent players' actions, ultimately creating a pseudo-turn-based mechanism.

Let's understand how to achieve that pseudo-turn-based mechanism:

1. Open the res://05.online-quiz/QuizScreenServer.gd script and let's implement its main methods.

2. First of all, let's add "any_peer" to the answered() method's @rpc annotation. This will allow any player to trigger the behavior we are about to describe when they answer a question correctly.

3. Inside the answered() method, we will tell quiz_panel to update the round's winner, making an RPC to the "update_winner" method and passing the player's name, which is stored in the database. This will update every peer's QuizPanel about the round's winner:

```
@rpc("any_peer")
func answered(user):
    quiz_panel.rpc("update_winner", database[user]
        ["name"])
```

4. Then, we start a local timer that should wait for enough time to allow players to digest the round's winner. We also make an RPC on wait_label so everyone's WaitLabel displays the correct waiting time as well:

```
@rpc("any_peer")
func answered(user):
    quiz_panel.rpc("update_winner", database[user]
        ["name"])
```

```
timer.start(turn_delay_in_seconds)
wait_label.rpc("wait", turn_delay_in_seconds)
```

5. Now let's do the same thing on the missed() method. But we will make an RPC to "player_ missed" instead:

```
@rpc("any_peer")
func missed(user):
    quiz_panel.rpc("player_missed", database[user]
        ["name"])
    timer.start(turn_delay_in_seconds)
    wait_label.rpc("wait", turn_delay_in_seconds)
```

With that, QuizScreenServer handles both game states when a player wins or loses the round. Using RPCs, we can update all peers about what's happening in the game and set them ready for the next round. But we haven't seen how this actually works yet. Coming next, let's see what happens in QuizPanel when we call the update_winner() and player_missed() methods.

Updating players about the round

In a multiplayer quiz game, it's crucial to keep all players in sync with the game state, especially when someone has already answered a question correctly. The QuizScreenServer Main node is responsible for updating the game state and informing all the connected players about what just happened in the current round. To achieve this, the QuizScreenServer Main node makes an RPC to all the peers' QuizPanels. The QuizPanel node on each player's side will update the game state locally and prevent any further interaction until the next round begins.

The implementation of these methods ensures that all players are on the same page, and there are no discrepancies in the game state between players. With this approach, we can provide a fair and consistent gaming experience for all the players in the game.

Open the res://05.online-quiz/QuizPanel.gd file and let's implement the update_ winner() and player_missed() methods, together with their auxiliary methods as well, such as lock_answers() and unlock_answers():

1. Find the update_winner() method and add the "call_local" option to its @rpc annotation. We do that because when we make this RPC on the server, it should also update its own QuizPanel node as well:

```
@rpc("call_local")
func update_winner(winner_name):
```

2. Then, inside the update_winner() method, update the question_label.text property to display a message with winner_name:

```
@rpc("call_local")
func update_winner(winner_name):
    question_label.text = "%s won the round!!" %
        winner_name
```

3. Finally, we will make a call to the lock_answers() method. This will make players wait for the next round, as we'll see soon:

```
@rpc("call_local")
func update_winner(winner_name):
    question_label.text = "%s won the round!!" %
        winner_name
    lock_answers()
```

4. We can do the exact same thing in the player_missed() method. But here, we'll display a different message, communicating that the player missed the answer:

```
@rpc("call_local")
func player_missed(loser_name):
    question_label.text = "%s missed the question!!" %
        loser_name
    lock_answers()
```

With that, we have our user interface updating the players about their peers' actions. If a player answered right, they will know it, if a player answered wrong, they will know it. Now is the time to get them ready for the next round. Let's look at the lock_answers() and unlock_answers() methods.

5. In the lock_answers() method, we are going to run through all AnswerButtons, which are children of the Answers node, and disable them. This way, players won't be able to interact with these buttons anymore, preventing them from answering the question:

```
func lock_answers():
    for answer in answer_container.get_children():
        answer.disabled = true
```

6. We do the opposite in the unlock_answers() method, toggling off the disabled property on each AnswerButton node:

```
func unlock_answers():
    for answer in answer_container.get_children():
        answer.disabled = false
```

This will prevent and allow player interactions with the available answers to the current question. We can use this same approach to create an actual turn-based system where players take turns trying to answer one question at a time. Here's a challenge for you, our network engineer. As an exercise, implement a turn-based system using the knowledge you just acquired. You have everything necessary at your disposal.

A turn-based system is a way of structuring gameplay where each player takes a turn to make their move, before passing control to the next player. This is in contrast to real-time gameplay, where all players are acting at the same time. Turn-based systems are often used in strategy games, where players need to carefully plan their moves.

To implement a turn-based system in your quiz game, you will need to modify the existing code to add a new layer of logic. One approach would be to create a queue of players, with each player taking their turn in order. When it's a player's turn, they are allowed to answer the question, while the other players are locked out. Once they have answered, their turn is over, and the next player in the queue takes their turn.

To create this system, you could turn `lock_answer()` and `unlock_answer()` into RPC methods and use the `rpc_id()` method to directly lock or unlock the players' answer options based on whether they are the current active player.

In the next section, we'll understand how we can take the fundamental mechanics of a quiz game and make them work in an online context. It's going to be the core of this chapter as we will see how we evaluate whether the player answered the question correctly and how we load new questions, ensuring that all peers are looking at the same question.

Turning local mechanics into remote gameplay

Now that we can manage players' interactions and communicate the game state to players, it's time to implement the core features of a quiz. Coming up, we are going to implement the questions and answers. For that, we are going to use a questions database where we store them with their possible answers and the correct answer index.

Here, we will see how we load and deserialize these questions into `QuizPanel`. On top of that, we'll also understand how we make use of RPCs to keep everyone in sync. And of course, we will also implement the logic behind both when players choose the correct and the incorrect answer.

When a player chooses an answer, we need to compare it with the correct answer index, and if it's correct, we should notify `QuizScreenServer` about the correct answer. We'll also need to make use of RPCs to keep everyone in sync regarding the current question and answer status.

Moreover, we need to implement the logic behind what happens when a player chooses the incorrect answer. We can use the same locking mechanism that we used previously to prevent players from answering if someone has already provided a valid answer. Once we handle the incorrect answer, we need to notify `QuizScreenServer` about the incorrect answer and move on to the next round.

By implementing all of these features, we can create a robust and engaging quiz game that can be played by multiple players simultaneously. By using the database to load questions, we can make the game dynamic and varied. And by using RPCs and locking mechanisms, we can ensure that the game runs smoothly and that everyone is on the same page.

Understanding the questions database

First of all, let's start by taking a look at our questions database. Open the file at res://05.online-quiz/QuizQuestions.json. It looks like this:

```
{
    "question_01":
        {
            "text": "Which of the following is not a
                Node?",
            "alternatives": ["Sprite2D", "Line2D",
                "Area3D", "PackedScene"],
            "correct_answer_index" : 3
        },
    "question_02":
        {
            "text": "Which of the following is an image
                file?",
            "alternatives": ["Bot.png", "Landscape.txt",
                "BeautifulTown.json", "UglyDuck.md"],
            "correct_answer_index" : 0
        },
    "question_03":
        {
            "text": "Which of the following is a sound
                file?",
            "alternatives": ["Scream.txt", "Blabla.ogg",
                "Laser.gd", "Music.svg"],
            "correct_answer_index": 1
        }
}
```

Notice that we represent each question as a *key* that is also a dictionary. Each question has three keys as well: "text", "alternatives", and "correct_answer_index". The "text" key is the actual question statement, "alternatives" is an array of possible answers that we will turn into AnswerButtons, and "correct_answer_index" is the index in the "alternatives" array of the correct answer.

Knowing that, you can go ahead and create some questions on your own. Keep in mind that, by default, we have four `AnswerButtons`, so try to provide four values in the `"alternatives"` key. Otherwise, you'd need to implement an `AnswerButtons` factory to dynamically create them based on how many answers we load from the question.

Loading and updating questions

Now, let's understand how this process works under the hood of `QuizPanel`. Open the script at `res://05.online-quiz/QuizPanel.gd` and find the `update_question()` method. The first thing you'll notice is that it has an `@rpc` annotation.

This is because we design it in such a way that it's the server who calls it and tells it which question to load. We will see that in a moment, but for now, let's implement this method's logic:

1. Create a variable called `question` and set it equal to the result of calling the `pop_at()` method on `available_questions` with the `new_question_index` argument passed in. With that, we'll remove the current question from the list of available questions and store it so we can use it moving on:

    ```
    func update_question(new_question_index):
        var question = available_questions.pop_at
            (new_question_index)
    ```

2. Check whether the question is not equal to null. Since the `pop_at()` method returns `null` when it can't find a value in the index provided, we check that to know if there are still questions that we didn't use yet, in other words, if `available_questions` is empty:

    ```
    if not question == null:
    ```

3. If the question we got is not null, set the `question_label.text` property to the `'text'` property of the question dictionary stored in the questions array. This is how we display the question's statement:

    ```
    question_label.text = questions[question]['text']
    ```

4. Create a variable called `correct_answer` and set its value to the `'correct_answer_index'` property of the question dictionary stored in the questions array. Doing that, we keep the correct answer stored so we can compare it when players answer:

    ```
    correct_answer = questions[question]
        ['correct_answer_index']
    ```

5. Loop through the numbers 0 through 3, inclusive, using the `range()` function and the `for` loop. For each iteration, create a variable called `alternative` and set it equal to `i`, which is the current element of the `'alternatives'` array stored in the question dictionary. Set the text of the current child node of the `answer_container` node to `alternative`. With that, we display the alternative's text on its respective `AnswerButton`:

```
for i in range(0, 4):
    var alternative = questions[question]
        ['alternatives'][i]
    answer_container.get_child(i).text =
        alternative
```

6. After loading the question and its answers, let's call the `unlock_answers()` function. This basically starts the current round, allowing players to interact with `QuizPanel` again:

```
unlock_answers()
```

7. If `question` is `null`, meaning we don't have any questions left to play the quiz, we need to loop through each child node of the `answer_container` node using the `for` loop. For each iteration, we'll set the text of the `question_label` node to `'No more questions'`:

```
else:
    for answer in answer_container.get_children():
        question_label.text = "No more questions"
```

8. Since we reached the end of our quiz match, we can call the `lock_answers()` function to prevent any further interactions:

```
lock_answers()
```

After these steps, the `update_question()` method should look like this:

```
@rpc
func update_question(new_question_index):
    var question = available_questions.pop_at
        (new_question_index)
    if not question == null:
        question_label.text = questions[question]['text']
        correct_answer = questions[question]
            ['correct_answer_index']
        for i in range(0, 4):
            var alternative = questions[question]
                ['alternatives'][i]
            answer_container.get_child(i).text =
                alternative
        unlock_answers()
```

```
    else:
        for answer in answer_container.get_children():
            question_label.text = "No more questions"
        lock_answers()
```

With that, we have our main quiz mechanism set up. We can pick one of the questions we have in our database and display it to our players. You can check the `_ready()` callback to understand how we load the questions in memory and assign them to the `available_questions` variable.

As mentioned before, we are going to focus on the essentials here. Talking about the essentials, we are still missing one mechanic, which is how we validate the answers, right? Find the `evaluate_answer()` method and let's implement that:

1. Inside the `evaluate_answer()` method, create a variable called `is_answer_correct` and set it equal to the comparison between the `answer_index` and `correct_answer` variables. This will check whether the given answer index matches the index of the correct answer:

    ```
    var is_answer_correct = answer_index == correct_answer
    ```

2. Emit a signal called `answered` with the `is_answer_correct` variable as an argument. This signal will be used by other parts of the quiz to tell whether the player's answer was correct or not:

    ```
    answered.emit(is_answer_correct)
    ```

    In the end, our `evaluate_answer()` method is quite simple and does just what we need to know whether the player answered the current question correctly:

    ```
    func evaluate_answer(answer_index):
        var is_answer_correct = answer_index == correct_answer
        answered.emit(is_answer_correct)
    ```

You may have noticed that the `evaluate_answer()` method isn't an RPC function, right? It essentially emits a signal that tells us whether the player's answer was correct or not. So how does the server manage that?

In the upcoming section, we will understand how this information is passed around between the client and the server implementations of our quiz.

Sending players' answers to the server

Now, let's understand the final piece of our mechanics and how it behaves in a multiplayer network. In the previous section, we ended up with an evaluation of the player's answer that led to the emission of the `answered` signal.

The `answered` signal needs to be handled in a specific way to ensure that all players are kept in sync and that the game state is consistent across all peers. When a player submits their answer, the `answered` signal is emitted, and the server updates all peers about it through an RPC call.

The `answered` signal and its associated methods are crucial for maintaining the integrity of the game state across all players in a multiplayer network. Without them, players might see different game states and have different experiences, which would make the game less enjoyable and potentially unfair.

In this section, we will understand how this signal propagates through the network and updates all peers about the players' answers.

Open the `res://05.online-quiz/QuizScreenClient.gd` script and you will notice that, right at the beginning, we have a callback to the `QuizPanel.answered` signal.

Let's implement this callback:

1. Inside the method's body, use an `if` statement to check whether `is_answer_correct` is true:

   ```
   func _on_quiz_panel_answered(is_answer_correct):
       if is_answer_correct:
   ```

2. If `is_answer_correct` is `true`, make an RPC call to the answered method on the server using the `rpc_id()` method. With that, the server will update all peers about the round winner:

   ```
   func _on_quiz_panel_answered(is_answer_correct):
       if is_answer_correct:
           rpc_id(
               get_multiplayer_authority(),
               "answered",
               AuthenticationCredentials.user
           )
   ```

3. If `is_answer_correct` is `false`, make an RPC call to the missed method on the server using the `rpc_id()` method. Finally, if the player chose the wrong answer, the server updates all peers about it:

   ```
       else:
           rpc_id(
               get_multiplayer_authority(),
               "missed",
               AuthenticationCredentials.user
           )
   ```

The whole `_on_quiz_panel_answered()` implementation should look like this:

```
func _on_quiz_panel_answered(is_answer_correct):
    if is_answer_correct:
        rpc_id(
            get_multiplayer_authority(),
            "answered",
            AuthenticationCredentials.user
        )
    else:
        rpc_id(
            get_multiplayer_authority(),
            "missed",
            AuthenticationCredentials.user
        )
```

With that, the client implementation will notify the server about the players' interactions. In turn, the server will update the game state and tell all peers to also update their game states to match the server's. After that, we have the missing network pieces in place and our online quiz game is ready. Feel free to test it out and experiment with more questions!

Summary

In this chapter, we saw how to create an online quiz game using the Godot Engine 4.0 Network API. We covered the fundamental concepts of online multiplayer game development, including client-server architecture, game synchronization, and player interactions. Using the quiz game, we saw how to feature dynamic loading of questions from a JSON database, and how to display the current players in a quiz match. We created a mechanism to prevent players from answering questions when another player has already provided one, creating a pseudo-turn-based mechanism. Finally, we saw how to manage players' interactions and communicate the game state to players and how to implement the logic behind both correct and incorrect answers, loading new questions one round after another until there are no more questions to display.

In the next chapter's projects, we will dive deeper into implementing a turn-based mechanism for our online quiz game. As we saw in this chapter, we can use a similar approach to what we did with the pseudo-turn-based mechanism, but with some modifications to make it a true turn-based system.

Additionally, we will explore how to pass information about players' turns, such as who is currently taking their turn, what happened during the opponent's turn, and more. We will also learn how to set up win and lose conditions and update peers about them, which will be essential for creating a sense of accomplishment and challenge in our game.

By the end of the next chapter, you will have a deeper understanding of the game development process, including how to create engaging online multiplayer game mechanics and implement them using the Godot Engine 4.0 Network API. See you there!

6

Building an Online Checkers Game

In this chapter, we will delve into the captivating realm of creating an online multiplayer checkers game. We will apply the knowledge and skills we have acquired throughout this book to develop an engaging and interactive gaming experience.

Checkers, a classic board game enjoyed by players of all ages, provides the perfect canvas to explore the complexity of online multiplayer game development. We will learn how to leverage the power of the Godot Engine and its versatile features to create a seamless multiplayer experience that will have players strategizing, competing, and enjoying the game together.

To facilitate the synchronization of the game state across multiple players, we will introduce a powerful tool called the `MultiplayerSynchronizer` node. This node will play a crucial role in updating the positions of the checkers pieces across the boards of all connected players. By using this node, we can ensure that each player's game view remains consistent and up to date, enhancing the overall multiplayer experience.

Throughout this chapter, we will cover essential concepts, such as client-server architecture, game synchronization, and player interactions, which are fundamental to the development of any multiplayer game. By understanding these concepts and applying them to our checkers game, we will create a robust and engaging multiplayer experience that will captivate players from around the world. The following diagram showcases the final project, where players are playing against each other online!

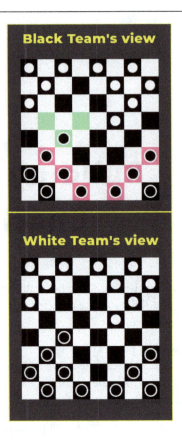

Figure 6.1 – Player 1's game instance view and Player 2's game instance view during the Black Team's turn

In addition to the technical aspects, we will also focus on creating an intuitive user experience that allows players to seamlessly navigate the game and interact with the checkers pieces. A well-designed interface is crucial in enhancing the player experience and ensuring a smooth and enjoyable gameplay session. Using some visual cues, we will ensure that players know the current game state intuitively.

As we progress, we will explore strategies to manage player interactions, such as handling player turns and validating moves. These features are vital in maintaining fairness and ensuring that the game follows the established rules of checkers. By incorporating these elements, we will create an authentic and immersive checkers experience that will engage players for hours on end.

By the end of this chapter, you will have gained valuable insights into the intricacies of online multiplayer game development. You will have the knowledge and skills necessary to create your own multiplayer games, with the ability to synchronize game states, handle player interactions, and deliver an immersive multiplayer experience.

So, get ready to embark on this exciting journey as we delve into the world of online multiplayer checkers game development. To do so, we will learn about the `MultiplayerSynchronizer` node and RPC functions, as they will be our key allies in synchronizing the players' boards. Let's begin this chapter and unlock the immense potential of multiplayer game development together.

We will cover the following topics in this chapter:

- Introducing the Checkers project
- Serializing players' turns
- Deserializing the opponent's turn
- Managing win and lose conditions

Technical requirements

In this chapter, we'll be working with the fourth folder of our Godot Engine project repository, which you can access through the following link: `https://github.com/PacktPublishing/The-Essential-Guide-to-Creating-Multiplayer-Games-with-Godot-4.0`.

Another requirement you will need to accomplish before following with our project import is to download Godot Engine version 4.0, as this is the version we will use throughout the whole book.

After opening your Godot Engine 4.0, open the project using the project manager. Then, navigate to the `06.building-online-checkers` folder. Here, you'll find all the files we used to build this chapter's project. You can test the game opening and playing the `res://06.building-online-checkers/CheckersGame.tscn` scene.

This scene showcases most of the features of our game. In this chapter, we are also going to implement the lobby system we have worked with throughout the book. On top of that, we will turn some local features into online multiplayer features as well.

So, stay tuned.

Introducing the Checkers project

Welcome, respected network engineer of our esteemed fictional indie development studio! As we embark on this chapter, let us take a moment to familiarize ourselves with the existing Checkers project. Currently, the project is designed for local multiplayers, allowing players to engage in thrilling matches offline. This understanding will serve as a solid foundation as we explore the path toward transforming our Checkers game into a captivating online multiplayer experience.

In this endeavor, our goal is to seamlessly transition the existing local multiplayer functionality into an online environment, without encountering significant obstacles along the way. By leveraging our existing knowledge and skills, we can effectively adapt the game to support online multiplayer, thus expanding its reach and providing players with the opportunity to compete with opponents from around the globe.

Throughout this section, we will unravel our Checkers project's inner workings, gaining valuable insights into its structure and mechanics. Armed with this understanding, we will be better equipped to navigate the porting process with confidence and efficiency.

As the designated network engineer, you play a pivotal role in this endeavor. Your expertise and problem-solving abilities will be put to the test. With careful consideration and strategic implementation, we can minimize any potential challenges and ensure a seamless transition to the online multiplayer realm.

Together, we will examine the project's architecture, dissect its components, and identify the necessary modifications required to facilitate online multiplayer functionality. By applying our knowledge of networking concepts and programming techniques, we will construct a robust foundation upon which the online multiplayer features will thrive.

Remember, you are an integral part of this process; as a network engineer, it is your job to implement all online multiplayer code, so you play a fundamental role here. Your skills will contribute to the realization of our vision – a thrilling online multiplayer Checkers game that captivates players across the digital landscape. So, let us embark on this journey. In the upcoming sections, we will understand the ins and outs of our local Checkers game.

Understanding the Checkers Piece scene

The **Piece** implements the functionality of a checkers game piece in our checkers' game. It allows players to interact with the Piece by selecting or deselecting it, keeping track of its selected state. In the following screenshot, we can see the SceneTree structure's Piece.

Figure 6.2 – The Piece scene node hierarchy

To enhance the player experience and decision-making during their turns, we utilize visual cues provided by the EnabledColorRect and SelectedColorRect nodes. EnabledColorRect becomes visible when the Piece is capable of making a valid move within a player's turn. Conversely, SelectedColorRect becomes visible when a player chooses this specific Piece over others, allowing for clear differentiation between them.

The Sprite2D node is responsible for displaying the current texture of the Piece. Depending on the player's team and whether the Piece has been promoted to a king piece, the texture can represent either a white piece, a black piece, or their respective king versions.

The SelectionArea2D node plays a crucial role in detecting player input. By using the input_ event signal, we establish communication between the player's clicks and the Piece. This enables us to toggle the Piece's selection state, determining whether it is currently selected or deselected.

Let's take a look at its code to understand how each of these nodes plays a role in this whole logic.

Overall, this code forms a crucial component of the online multiplayer checkers game, providing the necessary functionality to interact with and manage the behavior of individual checkers pieces within the game:

```
extends Node2D

signal selected
signal deselected

enum Teams{BLACK, WHITE}

@export var team: Teams = Teams.BLACK
@export var is_king = false: set = _set_is_king
@export var king_texture = preload("res://
    06.building-online-checkers/WhiteKing.svg")

@onready var area = $SelectionArea2D
@onready var selected_color_rect = $SelectedColorRect
@onready var enabled_color_rect = $EnabledColorRect
@onready var sprite = $Sprite2D

var is_selected = false
```

We start by defining some variables to store information about the Piece's team – whether it's a king, and its texture. It also has references to some child nodes in the scene and a variable to track whether the Piece is currently selected. Signals are defined to indicate when the Piece is selected or deselected. Next, we will declare the setter function we stated in the `_is_king` variable:

```
func _set_is_king(new_value):
  is_king = new_value
  if not is_inside_tree():
    await(ready)
  if is_king:
    sprite.texture = king_texture
```

When this variable is set to a new value, the code checks whether the node is part of the scene tree. If not, it waits until the node is ready, this prevents any errors when changing the variable's value through the inspector. If the variable is set to `true`, it updates the visual appearance of the game piece to represent a king, using a specific texture. This allows for dynamic visual changes in the game piece when it becomes a king. After that, we have a signal callback from the `Area2D` node's `input_event` signal:

```
func _on_area_2d_input_event(viewport, event, shape_idx):
  if event is InputEventMouseButton:
    if event.button_index == 1 and event.pressed:
      select()
```

It listens for left mouse button clicks and, when detected, calls the `select()` method, which, as we will see next, performs the procedures related to selecting the Piece. This code enables interaction within the game when the player clicks the left mouse button while hovering over the Piece's `Area2D`. Here, we have the code for the `select()` method:

```
func select():
  get_tree().call_group("selected", "deselect")
  add_to_group("selected")
  selected_color_rect.show()
  is_selected = true
  selected.emit()
```

In this function, we define what happens when the Piece is selected. It ensures that any previously selected objects are deselected, marks the current object as selected, shows a visual indicator, updates a variable to reflect the selection status, and emits the `selected` signal to notify other parts of the game about the selection event. This code is fundamental for managing and conveying the selected state of Pieces in the game. Then, we also have the opposite function, `deselect()`:

```
func deselect():
  remove_from_group("selected")
  selected_color_rect.hide()
```

```
    is_selected = false
    deselected.emit()
```

Here, we define what happens when an object is deselected. The code removes the object from the "selected" group, hides the visual indicator of selection, updates the is_selected variable to reflect the deselection status, and emits the deselected signal to notify other parts of the game about the deselection event. Now, it's time to enable the change in selection states; without that, the Piece shouldn't be selectable. This helps to prevent players from selecting Pieces from the opponent's team:

```
func enable():
  area.input_pickable = true
  enabled_color_rect.visible = true
```

In this part, we make the Area2D responsive to input events, and we also make a visual indicator visible on the screen. This is important to allow players to interact with and manipulate Pieces in the game when they are in an enabled state. Now, let's see its counterpart, the disable() method:

```
func disable():
  area.input_pickable = false
  enabled_color_rect.visible = false
```

It's essentially the opposite of the previous method. We make the Area2D non-responsive to input events, effectively making it non-interactive. On top of that, we hide the visual indicator on the screen that represents the Piece's availability, visually representing the Piece's disabled state. This is useful for controlling when and how players can interact with Pieces in the game.

In the upcoming section, we are going to see how the FreeCell scene works. It's a scene we use to highlight the available free cells that a selected Piece can move to.

Comprehending the FreeCell scene

In our game project, we have a concept called **FreeCell**, which represents a valid cell that a Piece can move to. Think of it as a designated area where the Piece is allowed to go. Each time the player selects a Piece with valid movement, we visually indicate the available cells by showing them in green. These cells are actually instances of the FreeCell scene, which we dynamically create and display on the game board.

To provide a clear example, imagine a scenario where a king Piece is selected. In the following diagram, you can see that all the cells where this king Piece can move to are highlighted in green. Each of these highlighted cells is an instance of the FreeCell scene, which allows the player to quickly identify the possible movement options for the selected Piece.

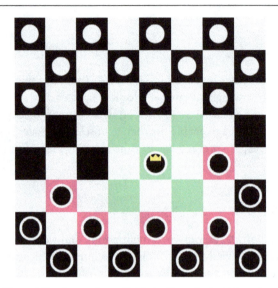

Figure 6.3 – Free cells available to the selected king piece

By using the FreeCell concept, we enhance the player's experience by visually communicating the valid movement possibilities. This empowers them to make informed decisions and strategize their next move effectively. It's a powerful tool that adds clarity and depth to the gameplay mechanics.

The FreeCell scene is an essential component of our game, composed of three distinct nodes. The main node is an `Area2D`, which serves as the foundation for the FreeCell. It encompasses two child nodes – a `CollisionShape2D` and a `ColorRect`.

The `CollisionShape2D` node is responsible for defining the shape and boundaries of the FreeCell. It ensures that the cell can properly interact with other game objects, such as Pieces or other elements within the game world. As we are about to see, `CollisionShape2D` also defines the boundaries in which the `Area2D` detects mouse inputs, which is fundamental for the FreeCell's behavior.

The `ColorRect` node, conversely, controls the visual representation of the FreeCell. It determines the color and appearance of the cell, providing a visual indicator to distinguish it from other elements on the game board.

Figure 6.4 – The FreeCell scene node hierarchy

To better understand the functionality and behavior of the FreeCell, let's explore its accompanying script. By examining the script code, we will gain insights into how the FreeCell operates and interacts with other game elements, ultimately contributing to the overall game logic and mechanics:

```
extends Area2D

signal selected(cell_position)

func _input_event(viewport, event, shape_idx):
    if event is InputEventMouseButton:
        if event.button_index == 1 and event.pressed:
            select()

func select():
    selected.emit(self.position)
```

This code allows the cell, represented by Area2D, to respond to right mouse button events within its area. When a cell is selected, it emits a signal with the position of the cell. This signal can be used to inform other objects or scripts about the selection and provide them with the cell's position for further processing. We'll use this in the CheckerBoard script to map the FreeCell's position into the board's cell.

In the upcoming section, we will introduce the game board, an essential component of our game. We will then shift our focus to the core of this chapter, which involves implementing methods that we will turn into RPCs. These RPCs will enable the game to work seamlessly in an online environment, allowing players to interact with each other and synchronize their actions over the network.

By leveraging the power of RPCs, we will create a dynamic and engaging multiplayer experience for our players. Let's dive into the details and explore how these methods will bring our game to life in an online setting.

Introducing the CheckerBoard scene

In this section, let's delve into the primary role of the game board, as this will serve as a foundation for our upcoming sections. By understanding the key responsibilities of the board, we can naturally identify the specific areas we will be focusing on in the subsequent sections. This understanding will pave the way for a clear and structured approach as we continue exploring our checker's development process.

The primary role of our game board is to manage the relationship between the black and white pieces in the game. To achieve this, the board utilizes the built-in TileMap functions to map the pieces onto cartesian coordinates. Additionally, it employs a hash map to associate the content of each cell on the board.

This means that we can access the content of a specific cell by providing its corresponding `Vector2i` coordinate. For example, by using the `meta_board[Vector2i(0,3)]` expression, we can retrieve the contents of the cell located in the first column and fourth row of the board. The result of this access will either be `null`, indicating that the cell is free, or it will return the Piece that is currently mapped to that particular cell. This mechanism allows for efficient retrieval and manipulation of the contents on the board, enabling seamless gameplay and interaction with the pieces.

The CheckerBoard plays a vital role in our game by overseeing various aspects of gameplay. Firstly, it manages the available movements for each cell on the board, taking into account the current team in play and whether a Piece has been promoted to a king. This ensures that players can only make valid moves based on the rules of the game.

The CheckerBoard is also responsible for controlling the turns in the game. It enables and disables the Pieces according to the active team, allowing only the team in play to make moves during their turn. This mechanism ensures fair gameplay and maintains the flow of the game.

Furthermore, the CheckerBoard keeps track of the number of Pieces each team possesses at the end of each turn. This count is crucial, as it determines the win-lose condition of the game. If a team has no remaining Pieces on the board, the CheckerBoard triggers the appropriate `win` condition, declaring the opposing team as the winner.

By managing the cell movements, regulating turns, and monitoring the Piece count, the CheckerBoard maintains the game's rules and progression. Its role is fundamental to providing clear conditions for victory or defeat. Let's take a look at the CheckerBoard scene structure:

Figure 6.5 – The CheckerBoard scene node hierarchy

It's important to note that the CheckerBoard is implemented as `TileMap`, a useful class in our game. We use specific methods provided by the `TileMap` class, such as `map_to_local()`, `local_to_map()`, and `get_used_cells()`, to establish our cell-mapping functionality. The `map_to_local()` method will help us map the game to cell positions in `TileMap`, while `local_to_map()` will help us convert the Pieces' positions to cells in the map. This will help us abstract the game in terms of rows and columns instead of floating-point numbers. As for the `get_used_cells()` method, it will help us access only the cells that have a tile set to them and avoid dealing with blank cells in `TileMap`. This will be useful when we create a matrix of the cells' contents.

In the `CheckerBoard` class, we'll focus on understanding the significance of the `meta_board` **Dictionary**, which serves as a fundamental component in subsequent sections. The core features we need to understand in order to turn this project into an online multiplayer checkers game are the `create_meta_board()` and `map_pieces()` methods within the `CheckerBoard` class:

```
func create_meta_board():
    for cell in get_used_cells(0):
        meta_board[cell] = null


func map_pieces(team):
    for piece in team.get_children():
        var piece_position = local_to_map(piece.position)
        meta_board[piece_position] = piece
        piece.selected.connect(_on_piece_selected.bind
            (piece))
```

The `create_meta_board()` method is responsible for setting up the `meta_board` Dictionary. This Dictionary acts as a data structure that maps cell coordinates to their corresponding contents. By leveraging the `TileMap` methods mentioned earlier, the `create_meta_board()` method populates `meta_board` with the appropriate cell coordinates and initializes them with null values, indicating empty cells.

Conversely, the `map_pieces()` method performs an essential role in updating the `meta_board` to reflect the current state of the game. This method iterates over all the Pieces on the provided team, which is passed as a reference to either the `BlackTeam` node or the `WhiteTeam` node. Then, it converts the Pieces' positions using the `TileMap.local_to_map()` method and maps each Piece to its respective cell coordinate in `meta_board`. This ensures that `meta_board` accurately represents the placement of Pieces on the visual board.

Lastly, the code establishes a connection between the `Piece.selected` signal and the `_on_piece_selected()` callback function. By connecting this signal, we bind the current Piece to the callback function as its argument. This enables us to conveniently access the `Piece` node whenever the player selects it.

This connection ensures that when the Piece emits the selected signal, the associated callback function, `_on_piece_selected()`, will be invoked and provided with the `Piece` node as its argument. This mechanism allows us to perform specific actions or access properties of the `Piece` node, in response to the player's selection.

By establishing this connection, we create a seamless interaction between the `Piece` node and the corresponding callback function, enhancing the flexibility and responsiveness of our game.

There are some auxiliary functions that help us calculate available cells and coordinate the Piece's movement; feel free to check them out and understand how we check for available cells, how we capture cells, and other gameplay features. In the next section, we will explore a different aspect of our checkers development journey.

We will focus on how we can package and transmit all the pertinent information about a player's turn across the network, ensuring that other players are promptly updated with the current state of the game board.

By understanding this process, we will be able to establish efficient communication between players, facilitating a seamless multiplayer experience. This functionality is essential for maintaining synchronization and enabling real-time gameplay in our online multiplayer game. Stay tuned as we dive into the intricacies of transmitting and updating the game state across the network.

Serializing players' turns

In *Chapter 2, Sending and Receiving Data*, we explored an essential technique to recreate the game state across multiple players in a network. By serializing the relevant data and transmitting it in small portions, we ensure efficient utilization of network bandwidth while maintaining synchronization among peers.

Developing an understanding of what information is crucial to replicate the game state among players involves mastering the concept of abstraction in game development. In our case, this primarily revolves around `meta_board`, which is an abstraction of the relevant metadata of our game, such as the positional data and king state of the Pieces and the empty cells in the board.

Additionally, we need to consider the availability of Pieces, depending on the players' turn. Fortunately, most other elements of the game can be managed locally without requiring network synchronization.

To simplify the process of synchronizing node properties across networked peers, I would like to introduce you to `MultiplayerSynchronizer`. This powerful node takes on the responsibility of automatically synchronizing properties across peers, relieving us from the tedious task of manual synchronization.

With `MultiplayerSynchronizer` in place, we can focus on developing the game's logic and let the node handle the efficient transmission of data among players.

Working with MultiplayerSynchronizer

`MultiplayerSynchronizer` plays a vital role by allowing us to effortlessly sync and share the state of nodes across multiple players, without writing any additional code. To begin utilizing this functionality, we will add a `MultiplayerSynchronizer` node to the Piece's scene. This will ensure consistency in the game state of each player. Let's dive into the process of integrating the `MultiplayerSynchronizer` and harnessing its capabilities.

Setting up MultiplayerSynchronizer in the Piece scene

Open the res://06.building-online-checkers/Piece.tscn scene and add a MultiplayerSynchronizer as a child of the Piece node. Then, we'll set up the properties we want to synchronize:

1. With the MultiplayerSychronizer node selected, in the bottom panel on the **Replication** tab, click the **Add property to sync** button.

Figure 6.6 – The Replication menu from the MultiplayerSynchronizer node

2. From the pop-up menu, select the **Piece** node.

Figure 6.7 – Selecting the Piece node from the Pick a node to synchronize pop-up menu

3. After selecting the **Piece** node, another pop-up menu will appear, asking you to select the property you want to sync. From there, select the **position** property.

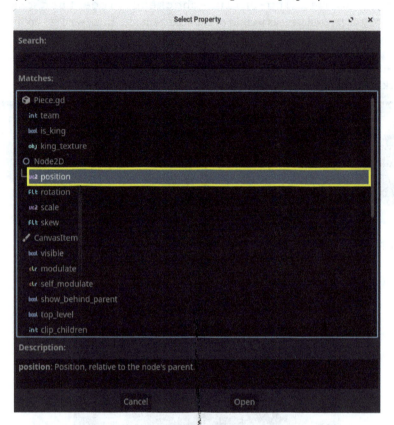

Figure 6.8 – Choosing the position property from the Select Property pop-up menu

And that's it. With that, as soon as the players are connected to the same network, they will automatically sync their Pieces' positions on the board. However, we still have an issue as this only updates the visual representation of the Piece, and we still need to update its data in meta_board. Now comes the fun part.

In the next section, we will start to understand the methods in the CheckerBoard that we need to turn into RPCs, keeping players on the same page.

Updating and syncing the CheckerBoard

A core issue when developing any application is that we have limited resources. In the case of transmitting data over a network, we are talking about bandwidth. Games are a special case because everything should happen in real time, so we can't risk heavy data transmission that hurts network performance.

To turn this in our favor, we need to pass data in the most abstract and lightweight way we can. In our case, we have our `meta_board` as a means to represent the game's current state. By using `Vector2i` coordinates, we can access and change the game states. And that's how we are going to keep players updated. In the next section, we are going to work with the `CheckerBoard.update_cells()` method, which is the core of our update system.

Using coordinates to update the board

Since `meta_board` is a Dictionary, we can access and set the values of its keys using Godot Engine's built-in types. For instance, if we want to change the content of the cell in the third row second column to `null`, we can write `meta_board[Vectori(1, 2)] = null`.

When Pieces perform a move, we just need to know the content of the previous and new cell of this movement, so that we can update it. And that's exactly what the `update_cells()` method does. Let's take a look at it:

```
func update_cells(previous_cell, target_cell):
    meta_board[target_cell] = meta_board[previous_cell]
    meta_board[previous_cell] = null
```

Since this is the very core of our update system, we need to turn it into an RPC function and call it as such.

For that, let's make the appropriate changes in our script:

1. Add the `@rpc` annotation to this method, using the `any_peer` and `call_local` options. We use these because we want every player to update others about changes in their board, and we also want their own board to update itself, hence the `call_local` option:

    ```
    @rpc("any_peer", "call_local")
    func update_cells(previous_cell, target_cell):
    ```

2. In the `move_selected_piece()` method, change the `update_cells(current_cell, target_cell)` line to an RPC call instead. This will make the call of this method both local and remote on other peers as well:

    ```
    func move_selected_piece(target_cell):
        var current_cell = local_to_map
            (selected_piece.position)
        Selected_piece.position = map_to_local
            (target_cell)
        rpc("update_cells", current_cell, target_cell)
        if not is_free_cell(target_cell):
            crown(target_cell)
    ```

With that, every time the CheckerBoard moves a Piece, it updates its `meta_board` data on all peers on the network.

Note that there's another method that we can turn into an RPC as well. Every player should update `Piece.is_king` of a Piece that reached their opponent's king row. For that, we have the `crown()` method that `move_selected_piece()` calls at the bottom of its logic.

Let's do the same thing we did with `update_cells()` with the `crown()` method:

1. First, we add the @rpc annotation to it with the `any_peer` and `call_local` options:

```
@rpc("any_peer", "call_local")
func crown(cell):
```

2. Then, we change the `crown(target_cell)` call to its `rpc()` version:

```
func move_selected_piece(target_cell):
    var current_cell = local_to_map
        (selected_piece.position)
    selected_piece.position = map_to_local
        (target_cell)
    rpc("update_cells", current_cell, target_cell)
    if not is_free_cell(target_cell):
        rpc("crown", target_cell)
```

With that, when a Piece reaches the king row, all players' CheckerBoards update their king state, be it an opponent Piece or an ally Piece.

Our work isn't done yet. In the next section, we will see how we update the `meta_board` content when players perform a capturing movement, meaning we need to remove a Piece from the board.

Removing a Piece from the board

When a player makes a move that ends up capturing an opponent Piece, we should update the game board accordingly. This means that on top of updating the cells involved in the movement, we should also update the cell where the captured Piece was, setting its content to null – in other words, turning it into a free cell. That's what the `remove_piece()` method does.

Let's take a look at its code:

```
func remove_piece(piece_cell):
    if not is_on_board(piece_cell):
        return
    if is_free_cell(piece_cell):
        return
    var piece = meta_board[piece_cell]
    piece.get_parent().remove_child(piece)
```

```
piece.free()
meta_board[piece_cell] = null
```

Since this behavior impacts both players, we need to turn this method into an RPC as well so that every time a player captures a Piece, they update their opponent with this sad fact.

Let's make the appropriate changes so that this feature is compliant with our online multiplayer demands:

1. Add the @rpc annotation to the remove_piece() method with the any_peer and call_local options:

    ```
    @rpc("any_peer", "call_local")
    func remove_piece(piece_cell):
    ```

2. In the capture_piece() method, update the remove_piece(cell) line to its rpc() version:

    ```
    func capture_pieces(target_cell):
        var origin_cell = local_to_map(selected_piece.
            position)
        var direction = Vector2(target_cell -origin_cell)
            .normalized()
        direction = Vector2i(direction.round())
        var cell = target_cell - direction

        if not is_on_board(cell):
            return
        if not is_free_cell(cell):
            rpc("remove_piece", cell)
            move_selected_piece(target_cell)
    ```

Now, every time a player captures a cell, it calls the remove_piece() method both locally and remotely on all connected peers!

With that, we have our players turn properly serialized and ready to be passed through the network to other peers, with good performance and little data usage, leaving a good bandwidth for us if we so desire. For instance, we can add a chat feature using a new RPC channel if we want to in the future.

In this section, we learned about the importance of abstracting relevant data for our network communication and how to turn local functionalities into remote functionalities, while maintaining all their logic and overall structure. Here, we saw the relevance of the call_local RPC option as well as the simplicity of turning a method call into an RPC call with the rpc() method.

In the next section, we will see how we manage the turn logic. This is an important feature to handle because there we will need to actively add a layer of network verification to properly handle the turns. The logic of a local turn shift and a remote turn shift is very distinct.

Handling remote turn shifts

One of the most important aspects of playing a game online is to maintain players' autonomy and authority over their resources – in this case, their team's Pieces. Godot Engine offers an interesting system where a SceneTree can structure its nodes' hierarchies with distinct Multiplayer Authorities.

To set up a node and its children's Multiplayer Authority, recursively, we can use `set_multiplayer_authority()` and pass the respective peer's ID as an argument. In our case, we are going to change the `BlackTeam` and `WhiteTeam` nodes' Multiplayer Authority to match their respective players' peer IDs.

This will be done by the server, so to keep the application simple, we are going to allow clients and server to share the same script, and we will check which one is running the server instance by using `is_multiplayer_authority()` on the CheckerBoard. We should only run this logic if the game is running in a network and there are peers connected. For that, we can check whether `multiplayer.get_peers().size()` is greater than 0, meaning there are peers connected. Let's see this in practice, shall we?

Setting up players' teams

The first thing we need to understand to handle players' turn shifts is that each one of the nodes that represent the team – in other words, the `BlackTeam` and the `WhiteTeam` nodes – should have its respective players' peer IDs set as their Multiplayer Authorities.

In that sense, we need to create a method in the `CheckerBoard` class that receives the team and the peer ID as arguments. Remember, we can't pass objects as arguments in this method because it needs to work in the network. So, we need to abstract teams as an `enum` that we can pass around through RPCs, and then all peers will be able to understand the message and access the correct team node at their end. Let's dive into the action and create a method called `setup_team()`:

1.　Add the `@rpc("authority", "call_local")` decorator before the `setup_team()` function definition. The `authority` option indicates that this RPC can only be called by the Multiplayer Authority; remember that the authority of the CheckerBoard will still be the server. The `call_local` argument specifies that the function should also be executed locally on the calling peers:

```
@rpc("authority", "call_local")
func setup_team(team, peer_id):
```

2. Inside the function, check whether the value of team is equal to Teams.BLACK; if this is the case, call the set_multiplayer_authority() method on the black_team object and pass peer_id as an argument. This effectively designates the specified peer as the authority for BlackTeam and all its children – in other words, the black Pieces:

```
@rpc("authority", "call_local")
func setup_team(team, peer_id):
    if team == Teams.BLACK:
        black_team.set_multiplayer_authority
            (peer_id)
```

3. Otherwise, call the set_multiplayer_authority() method on the white_team object and pass peer_id as an argument:

```
@rpc("authority", "call_local")
func setup_team(team, peer_id):
    if team == Teams.BLACK:
        black_team.set_multiplayer_authority
            (peer_id)
    else:
        white_team.set_multiplayer_authority
            (peer_id)
```

This method sets up the Multiplayer Authority of a team node based on the received team, black_team or white_team, using the provided peer_id. This ensures that the Multiplayer Authority for each team is correctly established, allowing the game logic to be synchronized across networked peers. Since the server calls this method on all peers, both players and the server will sync their team nodes accordingly.

Now, to ensure that this mechanism will be established among all peers, we are going to add the following lines of code right inside the _ready() callback:

1. Inside the _ready() callback, check whether there are peers connected in the multiplayer session by checking whether the size of multiplayer.get_peers() array is greater than 0:

```
func _ready():
    if multiplayer.get_peers().size() > 0:
```

2. If this is the case, check whether the current node is the Multiplayer Authority by using the is_multiplayer_authority() function. This ensures that we will only call the following logic in the server peer:

```
func _ready():
    if multiplayer.get_peers().size() > 0:
        if is_multiplayer_authority():
```

3. Then, make an RPC using the `rpc()` method, with the `"setup_team "`, `Teams.BLACK`, and `multiplayer.get_peers()[0]` arguments. This will call the `setup_team()` method on all connected peers, telling them to set the BlackTeam's Multiplayer Authority using the first peer ID in the list of connected peers. So, the first player connected in the session will be responsible for the black Pieces:

```
func _ready():
    if multiplayer.get_peers().size() > 0:
        if is_multiplayer_authority():
            rpc("setup_team", Teams.BLACK,
                multiplayer.get_peers()[0])
```

4. Right below the previous line, we are going to do the same thing but use `Teams.WHITE` and the connected peers' list second index, meaning the second player that connected to the session:

```
func _ready():
    if multiplayer.get_peers().size() > 0:
        if is_multiplayer_authority():
            rpc("setup_team", Teams.BLACK,
                multiplayer.get_peers()[0])
            rpc("setup_team", Teams.WHITE,
                multiplayer.get_peers()[1])
```

With that, we have our team setup in place. Note that, since the server has both team nodes' Multiplayer Authorities assigned to each of the players in the match, the server itself can't perform any movement in the board's Pieces.

Talking about that prevention mechanism, how does it work? How does the CheckerBoard prevent players from interacting with Pieces, especially with their opponent Pieces? In the next section, we are going to see how we can detect which player is assigned to which team and only re-enable their appropriate team Pieces.

Enabling and disabling team pieces

In our game, when players end a turn, we disable all their pieces using the `disable_pieces()` method. In turn shifts, we make sure to disable both teams' Pieces, and we also check whether there's a winner from the past turn; if not, we start the procedure to re-enable players' Pieces based on the turn's team.

All of that happens in the `toggle_turn()` method, but as it is, it won't work in an online multiplayer scenario because, currently, the method performs only local logic. So, let's turn it into a method that will work for our improved online multiplayer checkers game.

However, before that, let's see how the code is right now so we can already pick where we will need to make adjustments:

```
func toggle_turn():
    clear_free_cells()
    disable_pieces(white_team)
    disable_pieces(black_team)
    var winner = get_winner()
    if winner:
        player_won.emit(winner)
        return
    if current_turn == Teams.BLACK:
        current_turn = Teams.WHITE
        enable_pieces(white_team)
    else:
        current_turn = Teams.BLACK
        enable_pieces(black_team)
```

This function is responsible for managing the turn-based logic of a game. It first clears the available cells for movements and then disables the pieces of both teams. Then, it checks whether there is a winner and, if so, emits a signal to the CheckersGame script's _on_checker_board_player_won() method, indicating the winning team. If there is no winner, it switches the turn to the other team and enables the pieces of the corresponding team.

Can you point out where we need to make the necessary changes in order to make it work in our online version of the game? Remember that the game should work both locally and remotely, so we need to maintain the overall outcome of this method. Let's begin the process:

1. Decorate the toggle_turn() method with the @rpc annotation, using the any_peer and the call_local options. This indicates that any peer can call this method remotely in the multiplayer session, but they should also call it locally. This ensures that even if we are playing the game without joining a multiplayer session, we can call this method locally, using the rpc() method, and everything will still work:

   ```
   @rpc("any_peer", "call_local")
   func toggle_turn():
   ```

2. Inside that, check whether the current_turn is Teams.BLACK; we are going to move enable_pieces(white_team) inside yet another check. This time, we are going to check whether we don't have any peers connected, meaning we are playing the game alone or locally:

   ```
   if current_turn == Teams.BLACK:
       current_turn = Teams.WHITE
       if not multiplayer.get_peers().size() > 0:
           enable_pieces(white_team)
   ```

3. If we are not playing the game locally, we need to check whether the current player is the WhiteTeam's Multiplayer Authority, using the `multiplayer.get_unique_id()` method; if so, we can enable the **WhiteTeam** Pieces. And that's how we ensure that only the correct player will have their Pieces re-enabled:

```
if current_turn == Teams.BLACK:
        current_turn = Teams.WHITE
        if not multiplayer.get_peers().size() > 0:
            enable_pieces(white_team)
        elif white_team.get_multiplayer_authority()
            == multiplayer.get_unique_id():
            enable_pieces(white_team)
```

4. We are going to do the same thing, but inside the `else` statement, which handles whether the `current_turn` was from `Teams.WHITE`:

```
else:
        current_turn = Teams.BLACK
        if not multiplayer.get_peers().size() > 0:
            enable_pieces(black_team)
        elif black_team.get_multiplayer_authority()
            == multiplayer.get_unique_id():
            enable_pieces(black_team)
```

With that, every time we call the `toggle_turn()` method, we are going to check whether the peer has authority over the current Pieces in play, and we only allow them to select the Pieces from their team. Now, we still need to make a small change in order for this to be compliant with our network requirements. In the `_free_cell_selected()` callback, let's change the line that makes a direct call to the `toggle_turn()` method, making it a remote call using the `rpc()` method:

```
func _on_free_cell_selected(free_cell_position):
    var free_cell = local_to_map(free_cell_position)
    if can_capture(selected_piece):
        capture_pieces(free_cell)
    else:
        move_selected_piece(free_cell)
    rpc("toggle_turn")
    selected_piece.deselect()
```

Note that the `can_capture()` method is responsible for checking whether there's any enemy Piece around `selected_piece`, which can lead to a capture move. If this is the case, we call the `capture_pieces()` method, which will perform the capture movement on all possible enemy Pieces in the selected direction. Otherwise, if there's no capture movement available, we perform a simple movement calling the `move_selected_piece()` method, passing around `free_cell` as an argument.

Now, every time a player selects an available free cell to perform a move with a Piece, they will make a remote procedure call to `toggle_turn()`, telling all the connected peers to properly disable and re-enable their respective Pieces. Awesome, isn't it?

At this point, we have all the core mechanisms of our game in place, and we can play an online match with other players connected to our network. There's only one thing missing. We still need to communicate over the network when a player wins a match and allow players to play again.

In the next section, we are going to create a simple mechanism to allow players to rematch after one of them won the match for good.

Managing win and lose conditions

Excellent! We have successfully completed the development of the CheckerBoard scene, and our game's core functionalities are now in place. The next step is to transition the logic of the CheckersGame scene from local to remote gameplay.

To begin, let's open the `res://06.building-online-checkers/CheckersGame.tscn` file and familiarize ourselves with its structure.

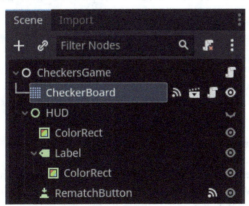

Figure 6.9 – The CheckersGame's scene node hierarchy

Take note that the CheckerBoard's `player_won` signal is connected to the `CheckersGame._on_checker_board_player_won()` callback. This callback is responsible for handling situations when a player's team has no remaining pieces on the board. Now, let's proceed by opening the script for CheckersGame.

We will be working on all the methods within the script, ensuring they are properly adjusted for online multiplayer functionality:

1. First of all, let's add the `@rpc` annotation to the `update_winner()` method with the `any_peer` and `call_local` options.

   ```
   @rpc("any_peer", "call_local")
   func update_winner(winner):
   ```

2. Then, we are going to do the same thing with the `rematch()` method. This one is called by RematchButton's `pressed` method:

   ```
   @rpc("any_peer", "call_local")
   func rematch():
   ```

3. Now, we need to call these methods remotely, using the `rpc()` method instead of directly calling them in CheckersGame. Let's do that in `_on_checker_board_player_won()`, turning the `update_winner(winner)` into `rpc("update_winner", winner)` instead. This is the method that the CheckerBoard's `player_won signals` is connected to:

   ```
   func _on_checker_board_player_won(winner):
       rpc("update_winner", winner)
   ```

4. Lastly, we do the same thing with `_on_rematch_button_pressed()`, turning the `rematch()` call into `rpc("rematch")`. This is the method that the `pressed` signal of **RematchButton** connects to, so when players press the button, this is what should happen:

   ```
   func _on_rematch_button_pressed():
       rpc("rematch")
   ```

With the adjustments we've made, our game is now fully equipped to run smoothly, whether it's played locally or remotely. When a player successfully captures their opponent's pieces, the game will transition all peers into a rematch state, where any peer can initiate a new match and start a fresh game. This ensures that players have the option to engage in continuous gameplay sessions without the need to exit and restart the game manually.

Figure 6.10 – The CheckersGame rematch screen

Our game is finally working! We have a fully functional checkers game that players can play online and challenge each other playing multiple matches.

Summary

To recap, in this chapter, we introduced the MultiplayerSynchronizer node to synchronize properties across a network, established the concept of abstraction for effective data transmission, utilized the @rpc annotations to enable multiplayer functionality, and learned how to assign and manage Multiplayer Authority to ensure player autonomy and resource protection.

In the upcoming chapter, we will see how to develop an online Pong game. There, we will cover the modifications necessary to turn the local game into an online multiplayer one, setting up online multiplayer paddles, syncing remote objects in real time, and coordinating the paddle's position. For that, will use the MultiplayerSynchronizer node with a bit more depth than we did in this chapter. Also, we will talk about the importance of maintaining a shared game world for players in action-based games, which is very different from turn-based games.

7
Developing an Online Pong Game

It's time to slowly get into some more complex aspects of making online multiplayer games. In *Chapter 6*, *Building an Online Checkers Game*, we saw how two players can share the same game world and see their actions have repercussions in the other players' game states. This happened with players taking turns, so we didn't have one of the most troublesome aspects of online multiplayer games involved: time.

In this chapter, we are going to start working with action games, which have hand-eye coordination and response time as their core features. We'll start with making a replica of one of the simplest physics-based games out there: Pong. Using the base project as a starting point, we will then turn it into an online multiplayer Pong game where each player controls one paddle and the Godot Engine high-level networking features will be responsible for keeping players in sync within the same game world.

We will cover the following topics in this chapter:

- Introducing the Pong project
- Setting up online multiplayer paddles
- Syncing remote objects

Technical requirements

For this chapter, we are going to use our repository of online projects, which can be found through the following link:

`https://github.com/PacktPublishing/The-Essential-Guide-to-Creating-Multiplayer-Games-with-Godot-4.0`

With the project opened in Godot Engine, open the `res://07.developing-online-pong` folder; everything we need for this chapter is there. That said, let's start by understanding how our Pong project works and what we need to do to turn it into an online multiplayer game. As stated throughout the previous chapter, we will also use Godot Engine version 4.0, so if you have other versions of the engine, please make sure you are using the correct one.

Introducing the Pong project

Welcome to yet another project for our fake indie game development studio, network engineer! This time, we need to make the onboarding of our next project.

We have a Pong game that we think we can turn into a competitive online multiplayer game with some leaderboards and all this cool stuff. Your core task here is to make its core features playable by two players through a network. Let's understand what we have currently so we can point out what you going to modify.

How the player paddles work

The players' paddles are the most important thing in our project. They are the only thing players actually control and, as such, they are the main way players interact with the game. By moving them, the players can bounce the ball off to the other player.

Let's take a brief look at the **Paddle** scene. For that, open the `res://07.developing-online-pong/Paddle.tscn` scene. Its scene tree structure looks like this:

Figure 7.1 – The Paddle scene's node hierarchy

Note that the paddle itself is a `Node2D` node, whereas the actual physical body is its child. This is a good way to abstract game entities. They have a physical body, but they aren't a physical body. This allows us to make more sense of them in more levels of abstraction. Now, let's take a look at its script:

```
extends Node2D

@export var speed = 500.0
```

```
@export var up_action = "move_up"
@export var down_action = "move_down"

@onready var body = $CharacterBody2D

func _physics_process(delta):
    body.move_and_slide()

func _unhandled_input(event):
    if event.is_action_pressed(up_action):
        body.velocity.y = -speed
    elif event.is_action_released(up_action):
        if Input.is_action_pressed(down_action):
            body.velocity.y = speed
        else:
            body.velocity.y = 0.0
    if event.is_action_pressed(down_action):
        body.velocity.y = speed
    elif event.is_action_released(down_action):
        if Input.is_action_pressed(up_action):
            body.velocity.y = -speed
        else:
            body.velocity.y = 0.0
```

This code allows the paddle to move up and down based on user input. The speed variable determines the movement speed, and the up_action and down_action variables represent the input actions for moving the paddle. The script handles input events and adjusts the character's velocity accordingly. The paddle moves at a constant speed or it stops if no key is pressed.

In the next section, let's see how the ball works. It's another core object in our game that we will need to do some work on in order to turn it into an online multiplayer game.

Understanding the Ball scene

The ball is the passive element of the game and it essentially bounces around when it hits another physical body – either the paddles, the ceiling, or the floor. The latter two are StaticBody2D, which use WorldBoundaryShape2D in their CollisionShape2D.

Let's take a look at the **Ball** scene. Open the `res://07.developing-online-pong/Ball.tscn` scene. The scene structure follows:

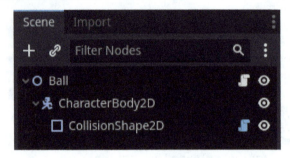

Figure 7.2 – The Ball scene's node hierarchy

Before we open the `Ball` script, notice that the `CollisionShape2D` Resource has a built-in tool script attached to it. It's a very simple script that draws the `CircleShape2D` Resource using the `CanvasItem.draw_circle()` method. This is the logic behind it:

```
@tool
extends CollisionShape2D

@export var color = Color.WHITE

func _draw():
    draw_circle(Vector2.ZERO, shape.radius, color)
```

That said, let's open the `Ball` script and see how it works, pay attention especially to the bouncing logic because it uses some interesting `Vector2` methods:

```
extends Node2D

@export var speed = 600.0

@onready var body = $CharacterBody2D

func move():
    body.velocity.x = [-speed, speed][randi()%2]
    body.velocity.y = [-speed, speed][randi()%2]

func reset():
    body.global_position = global_position
    move()

func _physics_process(delta):
```

```
    var collision = body.move_and_collide
        (body.velocity  delta)
    if collision:
        body.velocity = body.velocity.bounce
            (collision.get_normal())
```

This code moves the ball's `CharacterBody2D` node at a specified speed, randomizes its motion direction, detects collisions with other objects, and makes the character bounce off surfaces upon collision. By using this script, the ball performs a dynamic and responsive movement with collision detection.

In the next section, we will understand how we detect when a player scores against the other.

Managing players' scores

When the ball reaches the left or right of the screen, the player on the opposite side should score. To detect this condition, we make use of a `ScoreArea` node, which, in essence, is an `Area2D` node.

Open the `res://07.developing-online-pong/ScoreArea.tscn` scene and we'll take a look at its scene tree structure:

Figure 7.3 – The ScoreArea scene's node hierarchy

As mentioned previously, it's just an `Area2D` node; it doesn't even have a `CollisionShape2D` Resource because, in this case, it's more friendly to add it in the final scene so we can choose a specific `Shape2D` Resource for each `ScoreArea` node. Now, let's take a look at its code:

```
extends Area2D
signal scored(score)
@export var score = 0
func _on_body_entered(body):
    score += 1
    scored.emit(score)
```

This code provides a simple way to keep track of a score in a game using an `Area2D` node. It emits a scored signal whenever another physical body (in this case, the ball) enters the area, incrementing the score by one. By connecting to this signal, other game objects can respond to score updates and perform related actions. We use this signal later on in the `PongGame` class.

To ensure the `ScoreArea` node only detects the ball, we make use of **Collision Layers** and **Collision Masks**. On the ball, here's how these properties look:

Figure 7.4 – The Ball Collision Layer and Collision Mask properties

The ball is in the second physics layer, but it masks the first one. This is so it detects collisions with the paddle, the floor, and the ceiling. It needs to be only on the second physics layer because the `ScoreArea` node only masks the second layer. We do that to prevent the `ScoreArea` node from detecting any other physics body, for instance, the floor or ceiling:

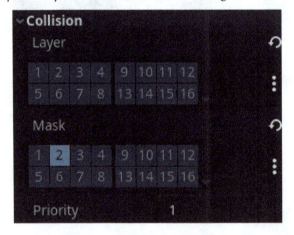

Figure 7.5 – The ScoreArea Collision Layer and Collision Mask properties

And that's how we ensure that the `ScoreArea` node will only interact with the ball. In the next section, we are going to see how we use this signal emitted by the `ScoreArea` node to actually update the displayed score, along with the overall logic behind the `PongGame` class.

Tying everything together

You may have noticed that these classes operate independently without any coupling, which means that they cannot form a cohesive system on their own. The responsibility of integrating everything into a coherent system falls on the `PongGame` class.

Let's first take a look at its scene tree structure so we can understand how everything will interact. Open the `res://07.developing-online-pong/PongGame.tscn` scene and pay attention to the **Scene** dock:

Figure 7.6 – The PongGame scene's node hierarchy

At this point, you already have a sense of most of these nodes. Let's just take a moment to understand what `ScoreLabel` nodes does. It's essentially just a text on the screen that displays each player's score. For that, it uses a method that changes its text property based on the casting of the score received by the `ScoreArea.scored` signal (which is an integer) into a string. The whole `ScoreLabel` node's code is as follows:

```
extends Label
func update_score(new_score):
    text = "%s" % new_score
```

With that in mind, let's jump into the `PongGame` code:

```
extends Node2D
@export var speed = 600.0
@onready var body = $CharacterBody2D
func move():
    body.velocity.x = [-speed, speed][randi()%2]
    body.velocity.y = [-speed, speed][randi()%2]
func reset():
    body.global_position = global_position
    move()
func _physics_process(delta):
    var collision = body.move_and_collide
        (body.velocity  delta)
    if collision:
        body.velocity = body.velocity.bounce
            (collision.get_normal())
```

This code keeps track of scores, displays a winner when a player reaches the target score, and allows the game to be restarted when one of the players presses the **Rematch** button, part of the `WinnerDisplay` Node interface. For visual reference, this is how `WinnerDisplay` node looks when toggled on:

Figure 7.7 – The WinnerDisplay overlay showing the match's winner

It also initializes the game by randomizing the ball's movement and starting its initial motion. Also, when a player scores, it resets the ball by recentering it and starting its movement again.

In this section, we went through all the core classes in our Pong game. They are currently meant for local multiplayer, so we need to modify them to support online multiplayer.

In the next section, we are going to do the necessary work to turn our game into a remotely playable Pong game where two players interact with each other, each with one paddle so they can compete together!

Setting up online multiplayer paddles

It's time to start your actual work. After understanding the whole project, let's do the necessary work to allow players to play it online!

In *Chapter 6*, *Building an Online Checkers Game*, we saw that changing the multiplayer authority of a `SceneTree` branch allows the new peer to take over control of the changes made to that branch of nodes. This was how we made it so that the player playing on the white team couldn't move the black team's pieces, and vice versa.

Being able to dynamically change the multiplayer authority is a core skill that we need to develop to maintain a coherent shared world for our players. In the situation we mentioned, players took turns in which each of them performed a single move and then the opposite player took control of their pieces. In this chapter, on the other hand, players must move simultaneously as this is an action game.

In the upcoming sections, we are going to implement a simple approach to give each player a paddle to play with.

Changing the paddle's owner

In our paddle implementation, we have a small issue to address. Both paddles call `CharacterBody2D.move_and_slide()` inside the `_physics_process()` callback, on top of checking for `InputEvent` in the `_unhandled_input()` callback. This makes it so that if the other player moves their paddle, the movement may be overwritten in the opponent's game. So, on top of re-assigning the paddles' multiplayer authority, we also need to disable the opponent's paddle callbacks. Open `res://07.developing-online-pong/Paddle.gd` and let's do it! Follow these steps:

1. Create a method called `setup_multiplayer()` and include the `player_id` argument, which represents the network identifier of the player:

    ```
    func setup_multiplayer(player_id):
    ```

2. Decorate the `setup_multiplayer()` function with the `@rpc` annotation and use the `call_local` option. This will ensure that only the server can call this method, and it will also call it locally on the server's endpoint:

```
@rpc("call_local")
func setup_multiplayer(player_id):
```

3. Inside the function, call the `set_multiplayer_authority()` method and pass it the `player_id` argument. With that, we have set up the paddle's new multiplayer authority. Now, we need to prevent its movement if the `player_id` argument doesn't match its multiplayer authority ID. We do that because this RPC function will be called on all peers, so the opponent's paddle is supposed to run the following code block:

```
@rpc("call_local")
func setup_multiplayer(player_id):
    set_multiplayer_authority(player_id)
```

4. Use the `is_multiplayer_authority()` method to check that the current paddle's peer ID doesn't match the multiplayer authority's peer ID:

```
        if not is_multiplayer_authority():
```

5. If this is the case, call the `set_physics_process()` function and pass it the `false` argument to disable physics processing:

```
        if not is_multiplayer_authority():
            set_physics_process(false)
```

6. Likewise, call the `set_process_unhandled_input()` function and pass `false` to disable handling unhandled input events on this paddle:

```
    if not is_multiplayer_authority():
        set_physics_process(false)
        set_process_unhandled_input(false)
```

7. At the end, the whole `setup_multiplayer()` method should look like this:

```
@rpc("call_local")
func setup_multiplayer(player_id):
    set_multiplayer_authority(player_id)
    if not is_multiplayer_authority():
        set_physics_process(false)
        set_process_unhandled_input(false)
```

This code sets up multiplayer functionality by assigning the paddle's multiplayer authority to a specified player. It then adjusts the behavior of the script based on whether the current instance is the authoritative peer. If the instance is not the authoritative peer, it disables physics and unhandled input processing to ensure that only the authoritative player performs those actions in this instance.

In the next section, let's understand how we gather and assign the player's ID to each respective paddle.

Assigning the players' paddles

Now that each paddle can have its own multiplayer authority and have independent physics and input handling processes for each player, it's time to understand how we are going to assign each player to their respective paddle. To do that, let's open the res://07.developing-online-pong/PongGame.gd script and, right in its _ready() function, let's create the necessary logic:

1. First of all, include the await keyword followed by the get_tree().create_timer(0.1).timeout expression. This creates a delay of 0.1 seconds and waits for its timeout signal to emit. This is important because we are going to use RPCs to call functions on remote nodes, and these nodes may not be ready by the time the game executes this code, so instead it waits a brief moment before executing its behavior:

```
func _ready():
    randomize()
    await(get_tree().create_timer(0.1).timeout)
    ball.move()
```

2. Then, check whether the size of the connected peers is greater than 0 by using the multiplayer.get_peers().size() method. This will ensure the following behavior only happens if there are peers connected; otherwise, the game runs as it should locally:

```
    await(get_tree().create_timer(0.1).timeout)
    if multiplayer.get_peers().size() > 0:
```

3. If this is the case, check whether the current instance is the current multiplayer authority by using is_multiplayer_authority(). This ensures that only the server will perform the player assignment:

```
    await(get_tree().create_timer(0.1).timeout)
    if multiplayer.get_peers().size() > 0:
        if is_multiplayer_authority():
```

4. Inside this condition, assign the first connected peer to the `player_1` variable. This will store the first player's ID in this variable:

```
await (get_tree().create_timer(0.1).timeout)
if multiplayer.get_peers().size() > 0:
    if is_multiplayer_authority():
        var player_1 = multiplayer.get_peers
            ()[0]
```

5. Then, assign the second connected peer to the `player_2` variable. This will store the second player's ID in this variable:

```
await (get_tree().create_timer(0.1).timeout)
if multiplayer.get_peers().size() > 0:
    if is_multiplayer_authority():
        var player_1 = multiplayer.get_peers
            ()[0]
        var player_2 = multiplayer.get_peers()
            [1]
```

6. Then, let's use the `rpc()` method to call the `setup_multiplayer` method remotely on the `player_1_paddle` and `player_2_paddle` nodes, passing their respective `player` variables:

```
if multiplayer.get_peers().size() > 0:
    if is_multiplayer_authority():
        var player_1 = multiplayer.get_peers()
            [0]
        var player_2 = multiplayer.get_peers()
            [1]
        player_1_paddle.rpc
            ("setup_multiplayer", player_1)
        player_2_paddle.rpc
            ("setup_multiplayer", player_2)
```

7. The whole `PongGame._ready()` callback should look like this after these changes:

```
func _ready():
    randomize()
    await (get_tree().create_timer(0.1).timeout)
    if multiplayer.get_peers().size() > 0:
        if is_multiplayer_authority():
```

```
var player_1 = multiplayer.get_peers()
    [0]
var player_2 = multiplayer.get_peers()
    [1]
player_1_paddle.rpc
    ("setup_multiplayer", player_1)
player_2_paddle.rpc
    ("setup_multiplayer", player_2)
ball.move()
```

This code demonstrates asynchronous programming and multiplayer setup. It starts by randomizing the random number generator and then introduces a delay of 0.1 seconds. It checks whether there are connected peers and whether the current instance is the multiplayer authority – in other words, the server. If these conditions are met, it assigns the connected peers to variables and uses RPCs to call the `Paddle.setup_multiplayer()` method with the respective peer information. Finally, it moves the `ball` object.

In order to properly set up who controls it, we provide `Paddle.setup_multiplayer()` with the required data – specifically, the player ID. However, a small problem arises when each player can only control their own paddle. How will players update their opponent's paddle position? Moreover, who should control the ball and how will its position be updated in both players' game instances? These questions will be addressed in the next section.

Syncing remote objects

With each player controlling their respective paddle in their own game instance, we have a small problem. The opponent's paddle will not update its movement because we ensured that both its physics and input processes were disabled after we assigned a new multiplayer authority. Due to that, the ball may also bounce on the opponent's paddle and create a different movement trajectory in each player's game instance. We need to ensure that players are sharing the same game world and, for that, we are going to sync the objects across the network using `MultiplayerSynchronizer`. In the next section, we are going to start the syncing of the ball.

Updating the ball's position

The first thing we are going to do is ensure the ball's position is synced across all peers. This is because we want to prevent players from dealing with different balls in their game instances, since this may lead them to make decisions based on wrong information. For instance, a player may move toward a ball that, in their game instance, is moving toward the ceiling, while in the server's game instance, the ball is actually moving toward the floor. Let's open `res://07.developing-online-pong/Ball.tscn` and start the simple process:

1. Add `MultiplayerSynchronizer` as the `Ball` node's child. We are going to use this node's features to keep all peers up to date with the ball's `CharacterBody2D` position:

Figure 7.8 – The Ball scene's node hierarchy with a newly added MultiplayerSynchronizer node

2. Then, using the **Replication** menu, let's choose the `CharacterBody2D:position` property to replicate across the connected peers:

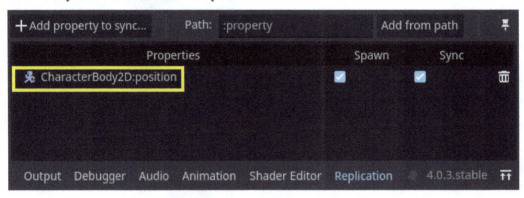

Figure 7.9 – The ball's CharacterBody2D position property in
the MultiplayerSynchronizer Replication menu

3. Finally, since we are working with physics bodies here, we need to ensure that the **Visibility Update** property is set to update during the **Physics** process. This will sync the `MultiplayerSynchronizer` updates to the local `Physics` update, ensuring that the game will take into account any collisions and other physics simulations when it updates the `Paddle` instances:

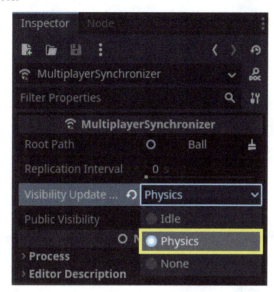

Figure 7.10 – The ball's MultiplayerSynchronizer's Visibility Update property set to Physics

4. To prevent any overwriting of the ball's position in any player's instance of the game, let's open the ball's script and add a code snippet at its `_ready()` callback stating that if this peer isn't the multiplayer authority, it will disable the ball's `_physics_process()` callback. This will make it so that only the server has the authority to actually calculate the ball's movement and, ultimately, its position, while players only replicate this in their game instances:

```
func _ready():
    if not is_multiplayer_authority():
        set_physics_process(false)
```

With that, the ball's movement should be the same across all connected peers, preventing them from making decisions based on an object that the other peers are seeing differently. This would break the game experience because, ultimately, the players would be playing in a different game world making movements that don't make sense to their peers. In the next section, let's do the same process for the `Paddle` object; of course, in this one, we won't need to disable `_physics_process()` because we do that when we set up its multiplayer authority.

Coordinating the paddle's position

Finally, it's time to sync the players' paddle positions to each other so they see their opponent's moves and can be on the same page. Let's open the res://07.developing-online-pong/Paddle.tscn scene and start the work:

1. Add a new MultiplayerSynchronizer as a Paddle child:

Figure 7.11 – The Paddle scene's node hierarchy with a newly added MultiplayerSynchronizer node

2. In the **Replication** menu, select the CharacterBody2D:position property to replicate across the connected peers:

Figure 7.12 – The Paddle CharacterBody2D position property in
the MultiplayerSynchronizer Replication menu

3. Just like in the **Ball** scene case, we are also working with a physics body here, so change the **Visibility Update** property to update during the **Physics** process:

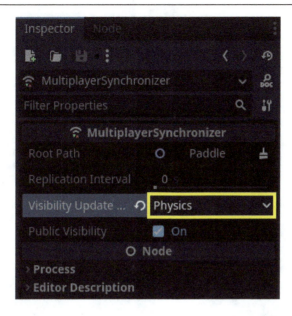

Figure 7.13 – The Paddle MultiplayerSynchronizer Visibility Update property set to Physics

With this implementation, each opponent's paddle will have its `CharacterBody2D` position synchronized across all peers in the game. This results in a shared game world that players can enjoy together while competing in a fair environment.

Summary

In this chapter, we learned how the Godot Engine High-Level Network API provides quick and easy solutions to assign the correct "owner" of a game object and sync its state across the network. This ensures that players are playing in a shared environment with an actual human opponent on the other side.

We learned how to check whether the current game instance is the multiplayer authority and make it perform the proper behavior accordingly. We also learned how to change the multiplayer authority of a node hierarchy on the `SceneTree`, ensuring that only a given player can make and sync changes regarding this node and its children. To sync the changes, we used `MultiplayerSynchronizer` with the **Physics** mode of **Visibility Update** to ensure that the physics interactions of the game objects are synced across all network peers.

In the upcoming chapter, we will strengthen our knowledge of online multiplayer games by creating a platformer game that two or more players can play together and explore the game world as they please. We are confident that this will be an exciting addition to our game development skills.

8
Creating an Online Co-Op Platformer Prototype

In this chapter, we will delve deeper into the work of creating action multiplayer online games. Our goal is to turn a local multiplayer puzzle platformer game prototype into an online version.

Here's what the final puzzle platformer prototype will look like:

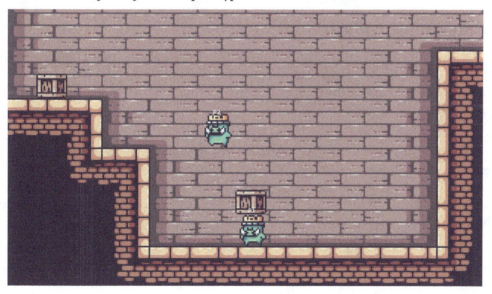

Figure 8.1 – A preview of the Puzzle Platformer prototype

By the end of this chapter, you'll know how to use the features of the `MultiplayerSpawner` node to create and assign playable characters to each player in the game, using the features offered by `MultiplayerSynchronizer` to sync relevant properties. With these features, we can go beyond updating the position of nodes; they will allow us to synchronize other properties, particularly animations. You will also learn how to leverage **Remote Procedure Calls** (**RPCs**) to manipulate the **multiplayer authority** of nodes. This will enable us to implement an exciting object-grabbing mechanic that will be a key element in our prototype.

Technical requirements

To access the resources for this chapter, you can find our repository of online projects by following the link provided here: `https://github.com/PacktPublishing/The-Essential-Guide-to-Creating-Multiplayer-Games-with-Godot-4.0`.

Once you have the repository, open the `res://08.designing-online-platformer` folder in the Godot Engine editor. All the necessary files for this chapter are located there.

Now, let's begin the onboarding process. In the upcoming section, we will familiarize ourselves with the project, explore its main classes, and identify where we need to implement networking features.

Introducing the platformer project

Our project presents a captivating puzzle platformer that will test players' strategic thinking and collaboration skills as they overcome challenging obstacles together. At the heart of this game lies a core mechanic centered around the manipulation of objects, utilizing them to construct platforms for the other player to traverse.

Now, let's dive into the essential classes that serve as the pillars of our project's foundation. Our first encounter will be with the `Player` class, which embodies the avatars controlled by each individual player. As the main protagonist, the `Player` class handles essential functionalities such as movement and interaction with various environmental elements. Notably, the `Player` class incorporates `InteractionArea2D` that detects contact with `InteractiveArea2D`, enabling players to perform specific actions upon them.

Moving forward, we encounter the `InteractiveArea2D` class. This class extends the functionality of the `Area2D` node and assumes the crucial role of a trigger area for detecting interactions. When `InteractionArea2D` overlaps with `InteractiveArea2D`, it becomes responsive to input events. Triggering the designated *interact* input action emits a signal, allowing us to create further engaging gameplay interactions.

In our game, the `Crate` class represents an interactive object that players can skillfully manipulate. Each `Crate` instance has an `InteractiveArea2D` node and a `CharacterBody2D` node, offering players the opportunity to collide with and leap onto them, which allows players to use them as viable platforms for navigating the level. These crates stand as major elements for puzzle-solving and advancing through the game's levels.

Lastly, we encounter the versatile `PlayerSpawner` class, responsible for the dynamic spawning and management of players within the game. This class adeptly adapts to the number of participating players, seamlessly instantiating a `Player` instance for each individual. Additionally, in the realm of local multiplayer, the `PlayerSpawner` class ensures a smooth and immersive gaming experience by establishing distinct controls for each player and optimizing gameplay customization.

In the upcoming section, we will dive into the `Player` object, which is composed of a script and a scene. We are going to understand how the script works with the available nodes on the scene and structures the desired behavior for our local `Player` node.

Understanding the Player class

The `Player` class and scene represent the player's avatar in the game. It is through this scene and script that players interact with the game world. The scene is a `CharacterBody2D` node with a `CollisionShape2D` Resource, a `Node2D` node called `Sprites`, which we use to group and pivot an `AnimatedSprite2D` node, and an `InteractionArea2D` node, which we are going to talk about in the *How the InteractiveArea2D class works* section. The `InteractionArea2D` node also has a `CollisionShape2D` Resource and a `RemoteTransformer2D` node, which we call `GrabberRemoteTransformer2D`. A `RemoteTransformer2D` node allows us to remotely sync the position, rotation, and scale of a node that is outside the hierarchy of the `RemoteTransformer2D`'s parent as if it were a sibling of the `RemoteTransformer2D`, which is very useful. In this case, we use the `GrabberRemoteTransformer2D` node to remotely transform the objects the player can grab, such as the *crate*, which we will talk about in the *Unveiling the Crate class* section. Finally, the `Player` class also has a `Label` node that we use to visually communicate the player controlling the avatar.

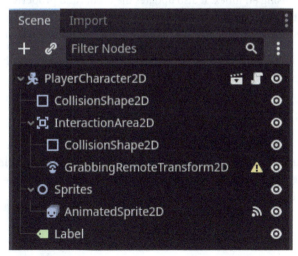

Figure 8.2 – The Player's scene node hierarchy

Now, let's look at the script itself. There are some aspects that we won't delve into in this section because they are more related to the basic platformer game player movement, and our focus here is on the online multiplayer aspect, so we would go beyond our scope. But the important part for your job is to know that when two players are playing locally, the `Player` node can set up its controllers dynamically so each player controls only one avatar. And this is something you'll have to make work in the online version of the prototype: how each player will control only their own avatar. For reference, the following code snippet does this locally:

```
func setup_controller(index):
    for action in InputMap.get_actions():
        var new_action = action + "%s" % index
        InputMap.add_action(new_action)
        for event in InputMap.action_get_events(action):
            var new_event = event.duplicate()
            new_event.device = index
            InputMap.action_add_event(new_action, event)
        for property in get_property_list():
            if not typeof(get(property.name)) ==
                TYPE_STRING:
                continue
            if get(property.name) == action:
                set(property.name, new_action)
```

The preceding code iterates over actions in `InputMap` singleton and creates new actions specific to a given controller device, using an index. It also updates the events and properties associated with the actions to be specific to the given device. The purpose of this code is to set up controller mappings for different players or devices in a game, allowing customization and differentiation of input controls.

In the next section, let's see how the *Crate* scene works, it's a pretty simple scene that essentially works as a passive object that players can use as a platform to move around the level.

Unveiling the Crate class

The `Crate` scene plays a crucial role in our game prototype. It represents interactive objects that players can skillfully manipulate to overcome obstacles and progress through the levels. Each instance of the `Crate` scene is a `Node2D` equipped with two important components: `InteractiveArea2D` node and `CharacterBody2D` node.

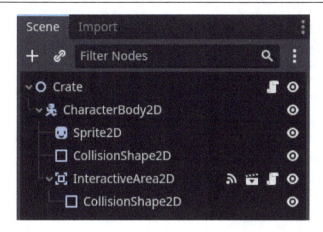

Figure 8.3 – The Crate's scene node hierarchy

The CharacterBody2D node represents the physical body of the *Crate* node within the game's physics simulation. It ensures that the crates collide with the player's avatar or other objects. The CharacterBody2D node handles the collision detection and response, allowing the player to seamlessly jump on and stand on the crate as if it were a solid platform.

As for InteractiveArea2D node, it is a special Area2D node that detects when InteractiveArea2D nodes overlap with it. In the context of the Crate class, the InteractiveArea2D node allows players to grab and lift the *Crate* nodes if they press the interact action while their InteractionArea2D node overlaps with the *Crate*'s InteractionArea2D node. This interaction enables players to use the *Crate* nodes as sturdy platforms to navigate the level since they will be able to move them around and jump on them even when another player is grabbing them. InteractiveArea2D node acts as a trigger, detecting when the player's avatar comes into contact with the Crate node and assigning the Crate node to the player's avatar GrabbingRemoteTransform2D node, synchronizing its position with the player even when they are moving.

The Crate script is fairly simple and structures how the Crate node responds and updates to interactions with the Player node:

```
extends Node2D

@onready var body = $CharacterBody2D
@onready var shape = $CharacterBody2D/CollisionShape2D
@onready var interactive_area = $CharacterBody2D/
    InteractiveArea2D

var lift_transformer = null
```

```
func _on_interactive_area_2d_area_entered(area):
    lift_transformer = area.get_node
        ("GrabbingRemoteTransform2D")

func _on_interactive_area_2d_interacted():
    lift_transformer.remote_path =
        lift_transformer.get_path_to(body)
```

The preceding code sets up references to nodes in the Crate's scene hierarchy. It also defines two callback functions that handle signals from the Crate node's InteractiveArea2D node. When an InteractionArea2D node enters the Crate node's InteractiveArea2D node, we presume it is the Player node interacting and we retrieve the Player node's "GrabbingRemoteTransform2D" node, assigning it to the lift_transformer variable.

When an interaction happens, the code assigns the lift_transformer.remote_path node to the path from lift_transformer variable to the Crate node's body. Remember, lift_transformer variable is a RemoteTransform2D node. This is how we allow the Player node's GrabbingRemoteTransform2D node to remotely transform the Crate node's CharacterBody2D Node position.

In the next section, we will understand how InteractiveArea2D node detects players' interactions with the *Crate* node and its role in our game.

How the InteractiveArea2D class works

In this section, we'll understand the role of a major scene that lies at the heart of our game's mechanics. Called InteractiveArea2D node, this scene plays a fundamental role in detecting and enabling player interactions with various objects in the game environment. InteractiveArea2D node enables us to turn any object into an object the player can interact with. For example, in our prototype, we use InteractiveArea2D node to allow the player to grab a Crate node and move it around.

The InteractiveArea2D scene, built upon the foundation of the Area2D node, serves as a fundamental component in our game. Its primary function is to detect and ease player interactions with objects, particularly within the player-crate interaction mechanic. Through the use of signals and input handling, the InteractiveArea2D scene ensures smooth gameplay interactions.

Figure 8.4 – The InteractiveArea2D's scene node hierarchy

One of the standout features of our game is the player-Crate node interaction mechanism, offering players the ability to manipulate interactive objects. The InteractiveArea2D scene serves as the catalyst for this interaction, acting as the gateway through which players can engage with the objects that populate the game world.

Using signals, the InteractiveArea2D scene establishes communication channels with other game objects and systems. Whenever a player successfully interacts with an object, InteractiveArea2D node emits the interacted signal. On top of that, the scene emits signals to indicate the availability or unavailability of interactions, allowing us to provide visual and auditory feedback to players.

To detect player input, the InteractiveArea2D scene uses an _unhandled_input callback. When players press the designated interact input action, it triggers the interacted signal, signaling that an interaction has occurred. This control scheme allows the players to interact with the game world.

Understanding the role of the InteractiveArea2D scene and its seamless integration with the player-Crate node interaction system is key. Now, it's time to dive into the code and unleash the full potential of this vital scene in our game:

```
class_name InteractiveArea2D
extends Area2D
signal interacted
signal interaction_available
signal interaction_unavailable

@export var interact_input_action = "interact"

func _ready():
    set_process_unhandled_input(false)

func _unhandled_input(event):
    if event.is_action_pressed(interact_input_action):
        interacted.emit()
        get_viewport().set_input_as_handled()

func _on_area_entered(_area):
    set_process_unhandled_input(true)
    interaction_available.emit()

func _on_area_exited(_area):
    set_process_unhandled_input(false)
    interaction_unavailable.emit()
```

The `InteractiveArea2D` script extends `Area2D` node and provides interaction functionality. It emits signals when an interaction occurs, when an interaction becomes available, and when an interaction becomes unavailable. It also handles unhandled input events to trigger interactions.

In the upcoming section and the final part of the onboarding, we will see how we create and insert `Player` instances in the game world dynamically based on how many players are playing.

Understanding the PlayerSpawner class

`PlayerSpawner` scene is another vital component in our game that handles the creation and positioning of `Player` instances. The `PlayerSpawner` class, based on the `Marker2D` node, follows the `Spawner` pattern, enabling us to dynamically generate `Player` instances in the game world.

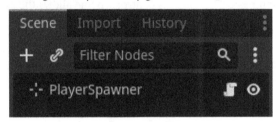

Figure 8.5 – The PlayerSpawner's scene node hierarchy

One of the key features of the `PlayerSpawner` class is its ability to position the spawned `Player` instances. As a `Marker2D` node, `PlayerSpawner` node provides a convenient way to specify the location and orientation of the players within the game world. This ensures that each player starts in the appropriate position, ready to embark on their adventure.

Let's see its code to understand what this class does under the hood:

```
extends Marker2D

@export var players_scene = preload("res://08.designing-
    online-platformer/Actors/Player/Player2D.tscn")

func _ready():
    if Input.get_connected_joypads().size() < 1:
        var player = players_scene.instantiate()
        add_child(player)
        return
    for i in Input.get_connected_joypads():
        var player = players_scene.instantiate()
        add_child(player)
        player.setup_controller(i)
```

The preceding script showcases the implementation of a `Spawner` class, based on the `Marker2D` node. It checks for connected joypads and creates instances of the `Player` scene accordingly. If no joypads are connected, it creates a single instance. If there are connected joypads, it creates one `Player` instance per joypad and sets up their respective controls. The preceding code snippet allows for the dynamic creation of `Player` instances in a multiplayer game, easing our work developing a multiplayer experience.

We are finally done with our onboarding; in the next section, we'll start to implement our online multiplayer features, turning our local prototype into something we can securely work with and polish knowing it's ready to launch with remote multiplayer features.

Spawning players in the match

In this section, we will understand how to improve the `PlayerSpawner` class to introduce online multiplayer features to our game. Leveraging the foundation laid by the *Understanding the PlayerSpawner class*, *Unveiling the Crate class*, and the *Understanding the Player class* sections, these enhancements enable multiple players to connect and interact seamlessly within a synchronized game environment.

`PlayerSpawner` node plays a fundamental role in our game's multiplayer architecture, acting as the core mechanism responsible for dynamically creating instances of the `Player` class for each connected player. These instances represent the avatars through which players engage with the game world.

With the integration of multiplayer functionalities, we will add features designed for an online multiplayer experience. This includes mechanisms to handle multiplayer authority, enabling proper gameplay across all connected peers. On top of that, the code will establish unique player names using the peers' IDs, allowing us to easily identify players on the network. To ensure synchronized actions, we will use RPCs, which will allow us to share events and actions among all connected players, especially the instantiation of other players.

One fundamental concept we are going to introduce here is the `MultiplayerSpawner` node. In the Godot Engine 4 High-Level Network API, the `MultiplayerSpawner` node is an invaluable asset for creating synced scenes in a networked multiplayer setting. In our context, it is a core component in synchronizing the creation of players, ensuring that every player can see and interact with the avatars of other players in real time.

With the `MultiplayerSpawner` node, we can effortlessly instantiate and position player avatars across all connected game instances. So to start with, let's open the `PlayerSpawner` scene at `res://08.designing-online-platformer/Levels/PlayerSpawner.tscn` and add a `MultiplayerSpawner` node as its child.

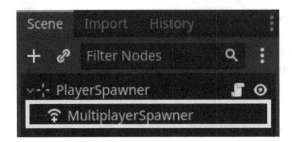

Figure 8.6 – A MultiplayerSpawner node as a child of the PlayerSpawner node

After that, we need to configure the `MultiplayerSpawner` node's **Spawn Path** and **Auto Spawn List** properties. The first property should point to `PlayerSpawner`. This tells `MultiplayerSpawner` who should be the spawned scenes' parent. Then, the second property should point to the same `PackedScene` Resource our `PlayerSpawner` node spawns. This will ensure that, when a new instance is created locally, `MultiplayerSpawner` node will replicate it on connected peers.

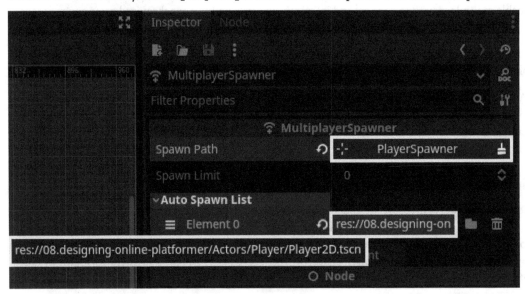

Figure 8.7 – The MultiplayerSpawner's Spawn Path and Auto Spawn List properties set up

With that, our `MultiplayerSpawner` node is ready to sync new players on everyone's game instances. But we still need to configure these new instances, otherwise, only the server will be able to control them. So, let's see how we can empower players with the ability to control their own avatars. Open the `PlayerSpawner` script at `res://08.designing-online-platformer/Levels/PlayerSpawner.gd`. In the next section, we will make some changes to this script.

Giving players control in PlayerSpawner

The new `PlayerSpawner` code introduces changes that enhance the multiplayer functionality of the game. Specifically, this code includes mechanisms to handle multiplayer synchronization and sets up `Player` instances correctly when multiple peers are connected. The changes involve checking for multiplayer authority, setting player names, and using RPCs to set up multiplayer functionality for each connected player. Let's implement these features:

1. Add `await(get_tree().create_timer(0.1).timeout)` at the beginning of the `_ready()` callback. This line introduces a delay of 0.1 seconds using a timer, allowing time for the multiplayer networking initialization to complete:

   ```
   func _ready():
       await(get_tree().create_timer(0.1).timeout)
   ```

2. Then, let's check whether there are connected peers by checking for the size of the `multiplayer.get_peers()` array. With that, we can check whether there are any connected peers in the multiplayer session. This condition verifies whether this is a local game session:

   ```
   func _ready():
       await(get_tree().create_timer(0.1).timeout)
       if multiplayer.get_peers().size() < 1:
   ```

3. If this is the case, we use the original logic we saw in the *Understanding the Player class* section to set up the local players' avatars' controllers. With a small twist, we use the `return` keyword at the end to prevent `_ready()` from reaching the next steps, which are only necessary if this is an online game session:

   ```
   func _ready():
       await(get_tree().create_timer(0.1).timeout)
       if multiplayer.get_peers().size() < 1:
           if Input.get_connected_joypads().size() < 1:
               var player = players_scene.instantiate()
               add_child(player)
               return
           for i in Input.get_connected_joypads():
               var player = players_scene.instantiate()
               add_child(player)
               player.setup_controller(i)
           return
   ```

4. Then, if this is an online game session, we check whether this game instance is the multiplayer authority (in other words, the server), and if so, we enter a loop that iterates over the connected peers:

```
if is_multiplayer_authority():
        for i in range(0, multiplayer.get_peers().
            size()):
```

5. Similar to the local session logic, we create a `Player` instance for each connected player:

```
if is_multiplayer_authority():
        for i in range(0, multiplayer.get_peers().
            size()):
            var player = players_scene.
                instantiate()
```

6. Here's the catch: after creating the `Player` instance, we set its name to the player's peer ID. Only then do we add it as a child of the `PlayerSpawner` node. This ensures each `Player` instance has a unique name and will prevent the RPCs and `MultiplayerSpawner` node from returning errors:

```
if is_multiplayer_authority():
        for i in range(0, multiplayer.get_peers().
            size()):
            var player = players_scene.instantiate()
            var player_id = multiplayer.get_peers()[i]
            player.name = str(player_id)
            add_child(player)
```

7. Then, we add another timer delay of `0.1` seconds. This delay gives time for peers' game instances to synchronize their multiplayer setup:

```
if is_multiplayer_authority():
        for i in range(0, multiplayer.get_peers().
            size()):
            var player = players_scene.instantiate()
            var player_id = multiplayer.get_peers()[i]
            player.name = str(player_id)
            add_child(player)
            await(get_tree().create_timer(0.1).
                timeout)
```

8. Finally, we make an RPC to the `Player.setup_multiplayer()` method passing `player_id` as an argument. `Player.setup_multiplayer()` is responsible for configuring the player's *multiplayer authority* based on the player ID, ultimately allowing this player, and only this player, to control this instance. We will implement this method in the *Setting up the Player multiplayer controls* section:

```
if is_multiplayer_authority():
        for i in range(0, multiplayer.get_peers().
            size()):
            var player = players_scene.instantiate()
            var player_id = multiplayer.get_peers()[i]
            player.name = str(player_id)
            add_child(player)
            await(get_tree().create_timer(0.1.
                timeout)
            player.rpc("setup_multiplayer", player_id)
```

We aren't done yet. We still need to set up the multiplayer features on other players' avatar instances when `MultiplayerSpawner` node creates them. For that, let's connect the `MultiplayerSpawner` node' spawned signal to the `PlayerSpawner` node using a method called `_on_multiplayer_spawner_spawned`.

Figure 8.8 – The MultiplayerSpawner spawned signal connecting to the
PlayerSpawner _on_multiplayer_spawner_spawned callback

Then, we make an RPC on the spawned node's `setup_multiplayer` method using the node's name as an argument. Since the name is a `StringName` variable, we need to convert it to a string and then to an integer in order for the `Player` class to handle it. The complete `PlayerSpawner` script should look like this after these changes:

```
extends Marker2D

@export var players_scene = preload("res://08.designing-
    online-platformer/Actors/Player/Player2D.tscn")

func _ready():
    await(get_tree().create_timer(0.1).timeout)
    if multiplayer.get_peers().size() < 1:
        if Input.get_connected_joypads().size() < 1:
            var player = players_scene.instantiate()
            add_child(player)
            return
        for i in Input.get_connected_joypads():
            var player = players_scene.instantiate()
            add_child(player)
            player.setup_controller(i)
        return
    if is_multiplayer_authority():
        for i in range(0, multiplayer.get_peers().size()):
            var player = players_scene.instantiate()
            var player_id = multiplayer.get_peers()[i]
            player.name = str(player_id)
            add_child(player)
            await(get_tree().create_timer(0.1).timeout)
            player.rpc("setup_multiplayer", player_id)

func _on_multiplayer_spawner_spawned(node):
    node.rpc("setup_multiplayer", int(str(node.name)))
```

The updated script incorporates multiplayer functionality by creating `Player` instances for each player in the network. It checks for the presence of connected joypads and multiplayer peers to determine the appropriate number of `Player` instances to create. The code also sets up the `Player` instances' controls and synchronizes their multiplayer settings. With these changes, `PlayerSpawner` node now enables multiplayer gameplay, allowing multiple players to control their avatars and interact within the game world simultaneously with no control conflicts.

In the upcoming section, we will explore the implementation of the `Player.setup_multiplayer()` method, which is responsible for configuring online multiplayer settings in the `Player` class. Within the `setup_multiplayer()` method, we set the multiplayer authority, disable the physics and input processing based on the local player's authority over the instance, and set a visual player index label updated with which player is controlling the instance.

Setting up the Player multiplayer controls

In this section, let's see how to implement the `Player.setup_multiplayer()` method, which plays a core role in setting up the online multiplayer controls for the `Player` class.

Within the `setup_multiplayer()` method, we need to take some key steps to achieve our online multiplayer controls. Firstly, we need to establish the new multiplayer authority, verifying the player's control and decision-making capabilities within the multiplayer environment. Then, we will adjust the physics and input processing based on whether the player ID matches the player ID we designated using the node's name. This ensures that each player controls the right `Player` instance.

On top of that, the method updates a visual player index label, allowing players to see their assigned avatar. This visual feedback enhances the multiplayer experience by providing a clear indication of each player's identity and presence in the game.

By implementing the `setup_multiplayer()` method, the game achieves synchronized multiplayer functionality, creating a cohesive and immersive multiplayer experience. Players can interact and collaborate with one another, encouraging a sense of shared adventure and enjoyment within the game world.

That said, let's dive into the code and unlock the potential of our multiplayer gameplay on our prototype! Open the `Player` script at `res://08.designing-online-platformer/Actors/Player/Player2D.gd` and let's implement the `setup_multiplayer()` method to finally allow players to control their avatars:

1. In the `Player` script, create a new method called `setup_multiplayer()`. It should receive an argument to get the player's ID; here, we'll call it `player_id`:

    ```
    func setup_multiplayer(player_id):
    ```

2. Then, decorate the method with the `@rpc` annotation, using the `"any_peer"` and `"call_local"` options. This specifies that the method can be called by any peer and executed locally. So, when players spawn their avatars, they tell the other peers to set up their avatars, setting up the avatar instance locally as well:

    ```
    @rpc("any_peer", "call_local")
    func setup_multiplayer(player_id):
    ```

3. Inside the `setup_multiplayer()` method, let's call `set_multiplayer_authority()` passing `player_id` as an argument to set the new multiplayer authority of this `Player` instance. Remember, the multiplayer authority determines the peer's control and decision-making capabilities over a given node:

```
@rpc("any_peer", "call_local")
func setup_multiplayer(player_id):
    set_multiplayer_authority(player_id)
```

4. Then, let's create a variable to store whether `player_id` is equal to the `Player` instance name. With that, we check whether the current avatar is supposed to be controlled by the local player:

```
@rpc("any_peer", "call_local")
func setup_multiplayer(player_id):
    set_multiplayer_authority(player_id)
    var is_player = str(player_id) == str(name)
```

5. After that, we set the physics and unhandled input processes based on the value of the `is_player` variable. With that, we disable the physics processing and the input handling on the `Player` instances that don't belong to the local player:

```
@rpc("any_peer", "call_local")
func setup_multiplayer(player_id):
    set_multiplayer_authority(player_id)
    var is_player = str(player_id) == str(name)
    set_physics_process(is_player)
    set_process_unhandled_input(is_player)
```

6. Finally, we update the text of the `label` node to display the player index. Here, `%s` is a placeholder that is replaced with the value returned by `get_index()`, representing the player's index in the `PlayerSpawner` children hierarchy (remember the first node is `MultiplayerSpawner`) so the player indexing starts at 1:

```
@rpc("any_peer", "call_local")
func setup_multiplayer(player_id):
    set_multiplayer_authority(player_id)
    var is_player = str(player_id) == str(name)
    set_physics_process(is_player)
    set_process_unhandled_input(is_player)
    label.text = "P%s" % get_index()
```

With that, we have our `Player` instance ready to behave in an online multiplayer environment. The `setup_multiplayer()` method configures the multiplayer features in the `Player` instances. It sets the multiplayer authority, adjusts physics processing and input handling based on the local player ID, and updates a label with the player's index.

But notice, since we are disabling physics and input processing, technically the other players' avatars will remain static during the whole gameplay session, right? Each player will only control and see their own character moving around and we don't want that. We want players to interact with each other and see how other players are behaving within this shared experience.

In the next section, we are going to use `MultiplayerSynchronizer` node to keep all other players on the same page regarding each other's avatar, including going beyond just the avatar's position, but also its animation and more. We will also see how we handle the `Crate` node: since players can grab and carry it around, who should have control over it? Who should be the `Crate` node's **Multiplayer Authority**?

Syncing physical objects

In this section, we will understand how to use the `MultiplayerSynchronizer` node for more than position updates. This node plays an important role in ensuring that players are synchronized with the avatars of other players in the game. As we have seen in the *Giving players control in PlayerSpawner* section, it is essential to maintain consistency among players to create a seamless multiplayer experience.

The `MultiplayerSynchronizer` class serves as a bridge between players, enabling real-time updates and synchronization of various properties. One key aspect that we will explore is how the `Crate` object's position is updated based on the player who is carrying it. This functionality allows for interactive and collaborative gameplay, where players can work together to solve puzzles or accomplish tasks.

On top of that, we will see how the `MultiplayerSynchronizer` node handles properties related to avatar animations. By leveraging the `MultiplayerSynchronizer` class, we can ensure that all players observe the same animation state of other players' avatars, enabling a visually consistent experience.

Through the usage of the `MultiplayerSynchronizer` node, we can establish a robust framework for synchronizing player actions, avatar positions, and animations. This synchronization ensures that all players perceive a cohesive and immersive multiplayer environment, fostering collaboration and enhancing the overall gameplay experience.

Let's explore the implementation details of `MultiplayerSynchronizer`!

Synchronizing the player's position and animations

Our `Player` scene has some nodes responsible for playing animations based on the player's actions and the current state of the avatar, namely the `Sprites` and `AnimatedSprite2D` nodes. It is important to synchronize the `Sprites` nodes' scale and the `AnimatedSprite` animation and frame since players' avatars would look rather weird if they jump, run, and stay idle and there's no visual feedback updating the performance of such actions in the game world other than the avatar's position changing. So, in this section, let's ensure that on top of the position, other relevant properties are also synced among players. For that, let's open the `Player` scene at `res://08.designing-online-platformer/Actors/Player/Player2D.tscn` and, of course, add `MultiplayerSynchronizer` as its child. With that, we will perform the following steps:

1. First of all, we need to change the `MultiplayerSynchronizer`'s **Visibility Update Mode** to **Physics** so it syncs the physics simulations on remote peers' game instances.

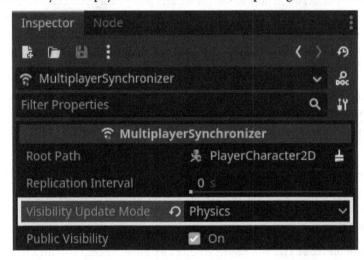

Figure 8.9 – The player's MultiplayerSynchronizer properties

2. After that, in the **Replication** dock, we are going to add the `PlayerCharacter2D` node's **Position** properties, the `AnimatedSprite2D` node's **Animation** and **Frame** properties, and the `Sprite` nodes' **Scale** properties. This will ensure that `MultiplayerSynchronizer` node also synchronizes the animation-related properties, allowing the players to see what their peers' avatars are doing.

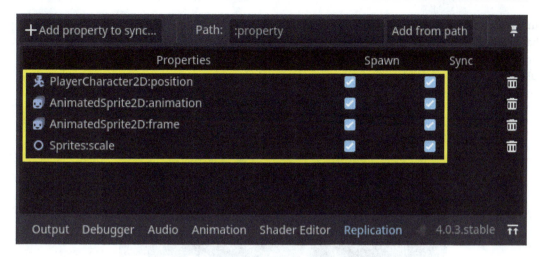

Figure 8.10 – The MultiplayerSynchronizer's Replication properties

And that's it! With that, our players are ready to interact within a shared game world. The `MultiplayerSynchronizer` node is an incredible ally to have in our toolbelt when developing online multiplayer games. As we can see in this section, those node allows for the synchronization of a range of different properties that can help us make our online gameplay experience enjoyable. There's a small, though very important observation to make in this regard. As we've seen throughout this book, especially in *Part 1*, we can't pass objects around, and we should avoid heavy data transmission through the network. So, keep that in mind when adding properties to the `MultiplayerSynchronizer` node's **Replication** menu. For instance, if you try to sync a **Texture** property, you are likely to fail the replication.

That said, in the next section, we are going to use `MultiplayerSynchronizer` node to sync the `Crate` node's position property, but there's a twist. Since any player can grab a `Crate` node and move it around, who should be its **Multiplayer Authority**? Well, that's what we are about to see!

Updating the crate's position remotely

At this point, we are fairly familiar with how `MultiplayerSynchronizer` node works and the overall concept of a node's multiplayer authority, right? One of the core mechanics in our online multiplayer puzzle platformer game is the ability for players to collaborate by taking objects and using them as platforms to progress through the levels' obstacles.

In this section, we are going to see how we can dynamically change an object's multiplayer authority based on which player is currently interacting with it so that only that player can change the object's properties. Open the `Crate` scene at `res://08.designing-online-platformer/Objects/Crate/Crate.tscn`, and add a new `MultiplayerSynchronizer` node as its child. Then, follow these steps:

1. Just like in the `Player` scene, we need to change the `MultiplayerSynchronizer` node's **Visibility Update Mode** to **Physics** to maintain the physics simulations consistently.

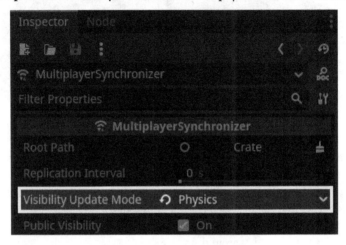

Figure 8.11 – The crate's MultiplerSynchronizer properties

2. Then, in the **Replication** menu, we are going to add the `CharacterBody2D` node's **Position** property to the syncing.

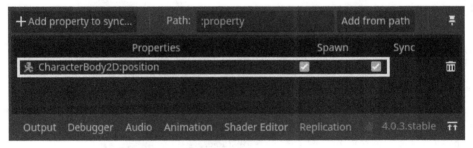

Figure 8.12 – The crate's CharacterBody2D position property in the Replication menu

With that, believe it or not, we already have what we need to sync the crate's position. Currently, the crate does not have any built-in behavior to move on its own, as its position is expected to be altered by the players who interact with it. To enable this functionality, we will make some additions to the Crate script. To get started, let's open the script file at res://08.designing-online-platformer/Objects/Crate/Crate.gd.

In the _on_interactive_area_2d_area_entered() method, we need to change the crate's multiplayer authority to match the player it's interacting with. For that, we can call the set_multiplayer_authority() method passing the area's multiplayer authority. This area that just entered is the player's InteractionArea2D node, so its multiplayer authority is the same as that of the player's:

```
func _on_interactive_area_2d_area_entered(area):
    lift_transformer = area.get_node
        ("GrabbingRemoteTransform2D")
    set_multiplayer_authority
        (area.get_multiplayer_authority())
```

With that, whenever the player's avatar enters the crate's InteractiveArea2D node, the player will become the crate's multiplayer authority and will be able to grab it and change its position once they interact with it. With this addition, we are ready to witness the seamless synchronization of the crate's position as players interact with it. You can test the prototype to explore the possibilities of collaborative gameplay and enjoy the immersive multiplayer experience we have just created!

Summary

In this chapter, we dived into the world of online multiplayer puzzle platformers, which emphasized teamwork and collaboration. Players will be challenged to work together, leveraging their skills to overcome obstacles and progress through intricate levels. Throughout the chapter, we explored key concepts and techniques to enhance the multiplayer experience and create a seamless collaborative gameplay environment.

To enable multiplayer functionality, we introduced the MultiplayerSpawner class, which dynamically instantiates Player instances based on the number of connected players. This ensures that each player has a unique avatar in the game, promoting a personalized and immersive multiplayer experience. The Player class played a crucial role, and we implemented the setup_multiplayer() method to configure its multiplayer settings. This method allowed us to set each instance's multiplayer authority, adjust physics and input processing, and update a visual player index label, providing players with a clear identification in the game.

To achieve synchronization between players, we harnessed the power of `MultiplayerSynchronizer` node. This powerful tool enabled us to synchronize not only the positions of players but also their animations. By incorporating `MultiplayerSynchronizer` node, we created a visually captivating multiplayer experience where players moved and interacted with the game world in perfect harmony. This synchronization brought the multiplayer gameplay to life, enhancing immersion and ensuring a cohesive and enjoyable shared experience.

An exciting feature we implemented was the ability for players to grab and manipulate the `Crate` object. By dynamically changing the crate's multiplayer authority, we ensured that only the player interacting with the crate had control over its movements. This added an extra layer of collaboration and puzzle-solving, as players can strategically use the crate as a platform to navigate the levels, fostering teamwork and coordination.

To sum up, this chapter provided a solid foundation for understanding and implementing multiplayer features using the Godot Engine High-level Network API. By combining the concepts and techniques explored, we created an online multiplayer puzzle platformer prototype, where players can seamlessly collaborate, synchronize their actions, and conquer challenges together. This chapter opened doors to endless possibilities in future multiplayer game development endeavors, empowering you to create engaging and interactive multiplayer experiences.

In the next chapter, we'll leverage all the knowledge we've seen so far in *Part 2* of this book to create a multiplayer online adventure with a persistent section system where players can log in and out and maintain their progress. The players will also synchronize the server's world with their game instance world, which also means they will be able to see all the other players that are currently playing as well and interact with one another. It's fundamentally a prototype that you can expand to a **Massive Multiplayer Online Role Playing Game (MMORPG)** if you want.

9

Creating an Online Adventure Prototype

In this chapter, we will explore the fascinating world of an online space adventure game that has the potential to evolve into a **massive multiplayer online role-playing game** (MMORPG). Throughout this journey, we will lay the foundations for an immersive gaming experience, allowing players to join a persistent world and seamlessly synchronize their game state with the current state of the game world.

Here's a snapshot of what the final spaceshooter adventure prototype will look like:

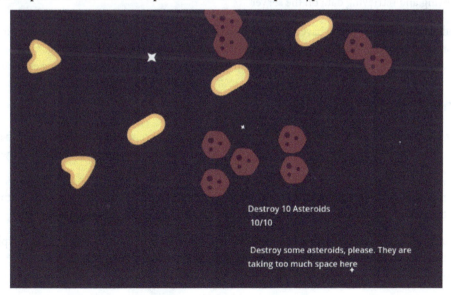

Figure 9.1 – Two players accomplishing the Destroy 10 Asteroids quest together

Our primary focus will be building a robust networking system that facilitates real-time interaction among players by using the powerful Godot Engine Network API. Through this system, players will connect to a central server, ensuring that everyone shares the same game world and can witness each other's actions, promoting collaboration and a sense of togetherness.

Furthermore, we will dive into the creation of a dynamic quest system capable of tracking player progress and storing this data in a database, so that when players come back, they will maintain their progress. Within our prototype of a space adventure, players will collaborate to complete missions such as destroying asteroids.

We will start the chapter by understanding the role of each piece of our game: the asteroids, the spaceship, and the player scene. Then, we will move to the core feature of an adventure game, the quest system, where we are going to learn how to pull and push data to the server and what builds up this robust system from both the server's and player's perspectives.

We will cover the following topics in this chapter:

- Introducing the prototype
- Logging the player in to the server
- Separating server and client responsibilities
- Storing and retrieving data on the server

By the end of this chapter, you will have a solid foundation for an online adventure game that can expand into a vast and captivating MMORPG. Equipped with a persistent world, synchronized gameplay, and a quest system, you will be well prepared to build an engaging and dynamic online gaming experience.

Technical requirements

To access the resources for this chapter, go to our online project's repository found at `https://github.com/PacktPublishing/The-Essential-Guide-to-Creating-Multiplayer-Games-with-Godot-4.0`.

With the repository in your computer, open the `res://09.prototyping-space-adventure` folder in the Godot Engine editor. You will find all the necessary files for this chapter there.

Now, let us begin the onboarding process. In the next section, we will introduce the project, explore its main classes, and identify where we need to implement networking features.

Introducing the prototype

In this section, we will gain a comprehensive understanding of the core systems driving our prototype. As the network engineers of our fictional studio, our role is core in transforming our local game prototype into an exciting online multiplayer game prototype. To accomplish this, we must familiarize ourselves with the major classes and files that make up our project.

Let's not overlook the significance of the onboarding process when we join a project. As network engineers, applying our knowledge and insights is essential for seamless integration into the development process. By understanding the core systems and concepts, we create a collaborative and productive environment, empowering the team to collectively bring our vision to life.

So, let's dive into the heart of our prototype and unlock the potential of online multiplayer gaming. By the end of this section, you will be equipped with the necessary systems you can tweak to shape an immersive and engaging online experience, uniting players in a dynamic and interconnected world. In the next section, let's understand how the `Player` class and scene work.

Understanding the Player scene

In any game, the player's avatar is a fundamental element of the player's experience. In this section, our goal is to understand the composition of the `Player` scene, the scene that represents the player's avatar in our prototype.

The `Player` scene is an abstract representation of the player as an entity in the game. It is a Node2D class that has a `Spaceship` scene, a `Weapon2D` scene, a `Sprite2D` node, a `HurtArea2D` node, a `CameraRemoteTransform2D` node, and, of course, the `Camera2D` node.

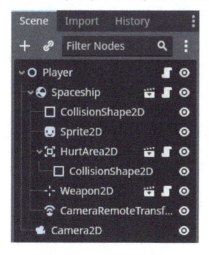

Figure 9.2 – The Player scene node's hierarchy

Now, let's understand the role of the main components of this scene, namely the Spaceship and Weapon2D nodes.

The Spaceship node is a direct child of Player node and carries most of the other components, with the exception of the Camera2D node; instead, it uses the CameraRemoteTransform2D node to remotely transform the Camera2D node. The Spaceship node is a RigidBody2D node that simulates the movement of a body within an environment with no gravity and very low friction. It has two main methods, Spaceship.thrust() and Spaceship.turn(). In the following code, we can see how we implemented these methods:

```
class_name Spaceship2D
extends RigidBody2D

@export var acceleration = 600.0
@export var turn_torque = 10.0

func thrust():
    var delta = get_physics_process_delta_time()
    linear_velocity += (acceleration * delta) *
        Vector2.RIGHT.rotated(rotation)

func turn(direction):
    var delta = get_physics_process_delta_time()
    angular_velocity += (direction * turn_torque) * delta
```

The thrust() method applies an acceleration force to Spaceship.linear_velocity property that makes it move. Then, the turn() method applies an acceleration force to the Spaceship. angular_velocity property that rotates it. We've set up the Spaceship node so that this is all it needs to perform a nice and smooth movement. It has some damping forces as well. In the following figure, we can see the properties related to the Spaceship scene to understand this movement better.

Figure 9.3 – The Spaceship RigidBody2D properties settings

The Player scene controls the Spaceship node's movement simply by calling the Spaceship. thrust() and Spaceship.turn() methods based on the player's inputs. In the following code snippet, we can see how this works in the _physics_process() callback:

```
extends Node2D

@export var thrust_action = "move_up"
```

```
@export var turn_left_action = "move_left"
@export var turn_right_action = "move_right"
@export var shoot_action = "shoot"

@onready var spaceship = $Spaceship
@onready var weapon = $Spaceship/Weapon2D

func _process(delta):
  if Input.is_action_pressed(shoot_action):
    weapon.fire()

func _physics_process(delta):
  if Input.is_action_pressed(thrust_action):
    spaceship.thrust()
  if Input.is_action_pressed(turn_left_action):
    spaceship.turn(-1)
  elif Input.is_action_pressed(turn_right_action):
    spaceship.turn(1)
```

Now, if you are an attentive engineer, which at this point we can assume you are, you might have noticed that in the _process() callback, we call the fire() method on the Weapon2D node, right? Let's understand how the Weapon2D node works; it's another core class for us.

The Weapon2D scene is a Marker2D node with BulletSpawner2D, Timer, Sprite2D, and AnimationPlayer nodes as its children. The following screenshot showcases the Weapon2D scene's structure:

Figure 9.4 – The Weapon2D scene's node hierarchy

BulletSpawner2D instantiates bullets and gives them a direction based on BulletSpawner2D's global_rotation value. We can see how this works in the following code block:

```
extends Spawner2D

func spawn(reference = spawn_scene):
  var bullet = super(reference)
  bullet.direction = Vector2.RIGHT.rotated(global_rotation)
```

As for Weapon2D, it uses Timer to establish a fire rate in which if Timer is currently active, it can't shoot. Otherwise, it plays the "fire" animation, spawns a bullet using whatever scene we set in its bullet_scene property, and starts Timer based on Weapon2D's fire_rate value. By default, it shoots three bullets per second. In the following code, we can see how we implemented this behavior:

```
class_name Weapon2D
extends Marker2D

@export var bullet_scene: PackedScene
@export_range(0, 1, 1, "or_greater") var fire_rate = 3

@onready var spawner = $BulletSpawner2D
@onready var timer = $Timer
@onready var animation_player = $AnimationPlayer

func fire():
  if timer.is_stopped():
    animation_player.play("fire")
    spawner.spawn(bullet_scene)
    timer.start(1.0 / fire_rate)
```

With that, players can shoot bullets and defend themselves while they accomplish their missions. One of these missions is to destroy some asteroids. So, in the next section, we are going to understand how the Asteroid scene works so we can move on to the quest system afterward.

Gauging the Asteroid scene

The Asteroid scene plays a fundamental role in our prototype. It represents an object that players must destroy in order to progress in a given mission. With the Asteroid scene working as planned, we can evaluate the quest system. In this section, we are going to understand how the Asteroid scene works so we have an idea of what to do in the process of turning the local gameplay prototype into an online multiplayer prototype.

The `Asteroid` scene is a `Node2D` node with an `AnimationPlayer` node, a `Sprite2D` node, a `GPUParticle2D` node, a `HitArea2D` scene, a `HurtArea2D` scene, a `StaticBody2D` node, and a `QuestProgress` scene. We can see the scene structure in the following screenshot:

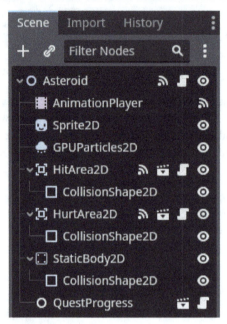

Figure 9.5 – The Asteroid scene's node hierarchy

`HitArea2D` node applies *1* damage to players' spaceships that touch it. When players shoot bullets, their `HitArea2D` node applies *1* damage to the `Asteroid` node if they hit its `HurtArea2D` node. If the `Asteroid` node doesn't have any hit points left, it plays the `"explode"` animation, emitting some particles through `GPUParticles2D` node and putting itself in the queue so `SceneTree` frees it from memory as soon as the animation finishes.

Doing so emits the `tree_exiting` signal, which is connected to `QuestProgress.increase_progress()` method. We are going to talk about `QuestProgress` node in the *Unraveling the quest system* section. The `Asteroid` node's behavior is expressed in the following code snippet:

```
extends Node2D

@export var max_health = 3
@onready var health = max_health

@onready var animator = $AnimationPlayer

func apply_damage(damage):
  health -= damage
```

```
    if health < 1:
      animator.play("explode")
    elif health > 0:
      animator.play("hit")

  func _on_hurt_area_2d_damage_taken(damage):
    apply_damage(damage)

  func _on_animation_player_animation_finished(anim_name):
    if anim_name == "explode":
      queue_free()
```

With that, the `Asteroid` node becomes a good testing subject for our quest system. In the next section, let's understand how this system works and the important aspects that we should consider for the online multiplayer version of our prototype.

Unraveling the quest system

It's time to understand the very core of what defines an adventure game. In this section, we are going to understand how the quest system of our prototype works and what we can do with it. This will enlighten us with a good understanding of what we need to change in order to turn it into a system that works for an online multiplayer version of the game.

Let's get started!

Representing a quest as a node

In this section, we will understand how the `Quest` node works and what we can do with it. To do that, open the `res://09.prototyping-space-adventure/Quests/Quest.tscn` scene. As with all other components of the quest system, it is a node with a script. Open the script and let's understand it.

The `Quest` node ultimately represents a quest in the player's quest log, and for that, it bundles all the data relevant to the quest itself:

- The `id` property that represents the quest in the database, `"asteroid_1"` by default.
- The quest's `title`, `description`, and `target_amount` properties. Note that these are exported variables, so this allows our (fake) quest designers to directly create new quests using the **Inspector**. You can see the **Inspector** displaying all these properties in the following figures:

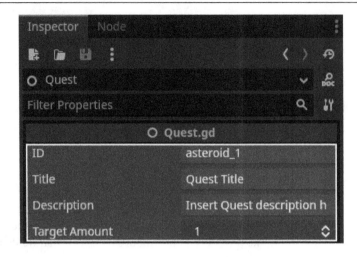

Figure 9.6 – The Quest properties in the Inspector

- It also has `current_amount` property to track the player's progress toward the quest's target amount, and a `completed` property to tell whether the player has finished the quest already or not.

- On top of that, it has a setter method for `current_amount` property to process the received value. It ensures that the value is clamped between 0 and `target_amount` property. It also emits a signal notifying the quest was updated, and if `current_amount` property is equal to `target_amount` property, it emits a signal notifying that the quest was completed.

In the following code snippet, we can see how this was implemented concretely:

```
extends Node

signal updated(quest_id, new_amount)
signal finished(quest_id)

@export var id = "asteroid_1"
@export var title = "Quest Title"
@export var description = "Insert Quest description here"
@export var target_amount = 1

var current_amount = 0 : set = set_current_amount
var completed = false

func set_current_amount(new_value):
  current_amount = new_value
  current_amount = clamp(current_amount, 0, target_amount)
```

```
    updated.emit(id, current_amount)
    if current_amount >= target_amount:
      finished.emit(id)
```

With that, we have an object that can represent a quest in our quest system. This is the very basic component of this system. As we saw, there is some data that makes up a `Quest`, right? This data is stored in a database so we can load and store the quest's content. In the next section, we are going to see what this database looks like and how we can load quest content and store any changes made to these quests.

Loading and storing quests with QuestDatabase

In any given adventure game, the players need quests to progress through the game story and overall world design. It is likely that these quests are stored in a place where quest designers can simply write them out and create an NPC to provide these quests to players.

Since in our fake quest designers would need some kind of database to design the quests, we made the `QuestDatabase` singleton. It loads JSON files containing all the available quests in the game and the player's progress in each of them. In this section, we will see how we can load these files and store the player's progress so they won't lose it when they leave the game, and how the `QuestDatabase` singleton provides this data to other classes.

Open the scene provided in the `res://09.prototyping-space-adventure/Quests/QuestDatabase.tscn` file and you will also notice that it's nothing more than a node with a script. In the **Inspector**, you'll notice the path to two important files:

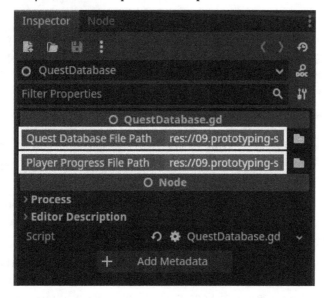

Figure 9.7 – QuestDatabase's Inspector

These are the JSON files that `QuestDatabase` uses to load the game's quest and the player's progress in the quests they have already started. The `PlayerProgress.json` file content is shown in the following code snippet:

```
{
    "asteroid_1": {
        "completed": false,
        "progress": 0
    }
}
```

So, each quest is represented by its ID and a dictionary with an indication of whether they are already completed or not and the progress the player has currently made. Now, for `QuestDatabase.json`, it is a bit more complex; the file content is shown in the following code snippet:

```
{
    "asteroid_1" : {
        "title": "Destroy 10 Asteroids",
        "description": "Destroy some asteroids,
            please. They are taking too much space here",
        "target_amount": 10}
}
```

Again, every quest is abstracted as a dictionary that reflects the quest's ID. Inside the dictionary, we have the `"title"`, `"description"`, and `"target_amount"` keys, which contain important data regarding the quest's object serialization and deserialization processes.

Now, the `QuestDatabase` singleton has some important methods to load, read, process, store, and even allow other objects to access this data. Let's briefly go through the main methods; you'll notice there are some extra methods in the class, but they are essentially there to retrieve specific information about a quest's data, such as the quest's title.

But let's focus on the more relevant methods:

- `QuestDatabase.load_database()`: Loads and deserializes the `QuestDatabase.json` and `PlayerProgress.json` files and stores their content, respectively, in the `quests_database` and `progress_database` member variables.

- `QuestDatabase.store_database()`: Does the opposite of the preceding method, serializing the `quests_database` and `progress_database` member variables into their respective files.

- `QuestDatabase.get_player_quests()`: Creates a `quest_data` dictionary for each key in the `progress_database` dictionary, gathering their data using the auxiliary methods, and returns a `quests` dictionary with all quests the player has started and their data.

- `QuestDatabase.update_player_progress()`: Updates the player's progress in a given quest. It receives a `quest_id`, `current_amount`, and `completed` argument to do so.

In the `QuestDatabase` script, we can see the concrete implementation of this behavior and the auxiliary methods. You will notice there's an implementation of the `_notification()` callback, which essentially calls the `store_database()` method when the application's window receives a close request:

```
func _notification(notification):
    if notification == NOTIFICATION_WM_CLOSE_REQUEST:
        store_database()
```

This guarantees that if the player quits the game through the usual means, for instance, clicking on the close button, their progress will be saved.

With that, we have the quest data and player's progress available at runtime and the quest system is almost done. We just need to know what we do with all that in the end, right? In the next section, we will understand how we use the intriguing `QuestProgress` nodes to update the system whenever a player makes progress in a given quest.

Managing players' quests

Now that we know that we can use the `QuestProgress` class when players get to progress in a given quest, we need to understand how these quests are managed in the system itself. In this section, we will understand how we retrieve quests from a quests database, how we create new quests for the current player based on the available quests retrieved, how we manage the player's progress in a given quest, and how we communicate that the player has a new quest in their quest log.

Open the scene available at `res://09.prototyping-space-adventure/Quests/QuestSingleton.tscn` and you will see it is a node with a script attached to it. Open the script and let's understand what this scene does.

As the singleton name, `Quests`, suggests, this scene is a set of all quests the player currently has. In the *Representing a quest as a node* section, we will see how we abstract each quest as an object with all the relevant properties, such as `title`, `description`, and `id`. The `QuestSingleton` class is responsible for retrieving and managing the quests.

To do that, it has three core methods:

- `QuestSingleton.retrieve_quests()`: Requests all the available quests for the players from the `QuestDatabase` singleton. We talked about `QuestDatabase` in the *Loading and storing quests with QuestDatabase* section.

- `QuestSingleton.create_quest()`: Receives a `quest_data` dictionary with all the relevant data to create a `Quest` node, then it instantiates a `Quest` and maps it to the `QuestSingleton.quests` dictionary using the quest ID. This allows other classes to access the `Quest` node using the quest ID in the upcoming method.

- `QuestSingleton.get_quest()`: Receives a `quest_id` value as an argument and uses it to return the given `Quest` node associated with the provided ID.

- `QuestSingleton.increase_quest_progress()`: Receives a `quest_id` value as an argument and an `amount` value to determine how much to increase in the provided quest's progress.

In the following code, we can see how these behaviors were implemented:

```
extends Node

signal quest_created(new_quest)
var quest_scene = preload("res://09.prototyping-space-
    adventure/Quests/Quest.tscn")
var quests = {}

func retrieve_quests():
  var player_quests = QuestDatabase.get_player_quests()
  for quest in player_quests:
    create_quest(player_quests[quest])

func create_quest(quest_data):
  var quest = quest_scene.instantiate()
  quest.id = quest_data["id"]
  quest.title = quest_data["title"]
  quest.description = quest_data["description"]
  quest.target_amount = quest_data["target_amount"]
  quest.current_amount = quest_data["current_amount"]
  quest.completed = quest_data["completed"]
  add_child(quest)
  quests[quest.id] = quest
  quest_created.emit(quest)

func get_quest(quest_id):
  return quests[quest_id]

func increase_quest_progress(quest_id, amount):
  var quest = quests[quest_id]
```

```
quest.current_amount += amount
QuestDatabase.update_player_progress(quest_id,
    quest.current_amount, quest.completed)
```

With that, `QuestSingleton` is able to retrieve all the quests the player is currently engaged in and provide them to user classes so they can access and work with said quests. This will allow us to actually increase the players' progress in a given quest. For that, we will understand how the `QuestProgress` node works.

Increasing quests' progress

Back in the `Asteroid` scene, we have the `QuestProgress` node. This node is responsible for communicating to the quest system when the player makes progress in a given quest. To know which quest `QuestProgress` refers to, we use a variable called `quest_id`, and this is a fundamental concept in our quest system. Through this data, other classes of the system can communicate with each other, requesting changes or retrieving information about a given quest.

On top of that, the `QuestProgress` class has a method called `increase_quest_progress()`, which requests `QuestSingleton`, referred to as `Quests`, to increase the quest's progress by the `amount` value provided, by default `1`.

We saw how `QuestSingleton` works in the *Managing players' quests* section. Nonetheless, in the following code snippet, we can see the `QuestProgress` class' code:

```
extends Node

@export var quest_id = "asteroid_1"

func increase_progress(amount = 1):
  Quests.increase_quest_progress(quest_id, amount)
```

The`QuestProgress` node itself is a small component of the system that works on the very end of the system, where the final output is processed. It is meant to be used by other classes to trigger its behavior. For instance, as mentioned in the *Gauging the Asteroid scene* section, the `Asteroid` node uses its `tree_exiting` signal to trigger the `QuestProgress.increase_progress()` method.

This concludes our quest system onboarding. Throughout this section, we understood how objects can increase a quest's progress, how we retrieve quests and players' progress from a database, how and what kind of data we store in the database files, and how this data ends up in a node in which we can implement higher-level behavior.

Our onboarding process is not over yet. In the upcoming section, we will understand how players will actually see the quest information in `QuestPanel` node, which is a component of the last piece of our prototype, the `World` scene. It is in this scene that all the action actually happens, so stay focused, and let's see how this works.

Breaking down the World scene

Everything we've seen so far will come together into the World scene. This is the scene where everything is put together to interact. It's the World scene that we use to test our current prototype. To do so, open the res://09.prototyping-space-adventure/Levels/World.tscn scene and hit the **Run Current Scene** button. You will be able to test the game and get a feel for the prototype.

Now, in this section, we are going to understand how we create asteroids and players in the game, and how we display the player's quest log on the screen. The world itself is a high-level abstraction scene, so things are easier to understand.

The scene itself is a node called Main that has a RadialSpawner child node called Asteroids responsible for spawning asteroids around it, a Spawner node called Players responsible for spawning Player instances, and some CanvasLayers nodes to create the overall visual of the game, namely BackgroundLayer node, which uses a ColorRect node to set the game's background color, then ParallaxBackground node, which has a ParallaxLayer node containing a GPUParticles2D node that creates a repeating starfield for the background.

Finally, we also have InterfaceCanvasLayer node, which, as the name suggests, contains interface elements. Here, we have an important element to wrap up the quest system: QuestPanel node. We are going to talk about it in the *Displaying quest information* section. In the following screenshot, we can see the World scene node hierarchy:

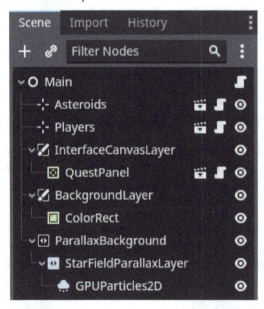

Figure 9.8 – The World scene node's hierarchy

Here, we were able to structure a prototype that spawns some Asteroids around a given area, spawns a pPlayer, and displays the player's quest log with the current active quests and their information. In the next section, let's understand how the QuestPanel node gathers, displays, and updates information about the player's quests.

Displaying quest information

In the end, the quest system has a main responsibility that summarizes everything we've seen so far regarding it. It has to display information about the current active quests to the player. This funnels down to QuestPanel node, which is a UI element that displays such information based on the data it gathers from QuestSingleton node. In this section, we are going to understand how QuestPanel node works. To do so, open the res://09.prototyping-space-adventure/ Quests/QuestPanel.tscn scene.

Note that QuestPanel node itself extends the ScrollContainer class and it has a VBoxContainer node as its child. This allows us to display many quests for the player and they will be able to navigate these quests using a scrollbar. We currently have only one quest, as seen in the QuestDatabase. json file, but the ground is paved for more quests. Now, open the QuestPanel script, and let's see how it implements displaying quest information.

The first thing it does right at the _ready() callback is to connect the Quests singleton's quest_created signal to QuestPanel's add_quest() method. Then it tells the Quests singleton to retrieve quests, which will populate the Quests singleton with the players' quests. Every time the Quests singleton creates a new Quest node, adding it as its child, it emits a signal that the QuestPanel node listens to and calls the add_quest() method. Let's talk about the QuestPanel node member variables and methods:

- quests_labels is a dictionary used to map Label nodes to their reference using the Quest.id property as the key.

- The add_quest() method creates a new Label node and sets its text property to a formatted String using the information from the Quest node stored in the quest property. It also connects the quest.updated signal to its update_quest() method, which we are going to talk about in a moment. Then, it adds this Label node as a child of theVBoxContainer node and maps it in the quests_labels property for further reference.

- The update_quest() method takes the quest_id String and the current_amount integer as arguments, and uses the quest_id argument to find the proper Label node to update the text with the updated quest data.

This behavior is expressed in the following code snippet if you want to understand the concrete implementation of how this all happens:

```
extends ScrollContainer

var quest_labels = {}

func _ready():
  Quests.quest_created.connect(add_quest)
  Quests.retrieve_quests()

func add_quest(quest):
  var label = Label.new()
  var quest_data = "%s \n %s/%s \n \n %s" %[quest.title,
      quest.current_amount, quest.target_amount,
          quest.description]
  label.autowrap_mode = TextServer.AUTOWRAP_WORD_SMART
  label.text = quest_data
  $VBoxContainer.add_child(label)
  quest.updated.connect(update_quest)

  quest_labels[quest.id] = label

func update_quest(quest_id, current_amount):
  var quest = Quests.get_quest(quest_id)
  var quest_data_text = "%s \n %s/%s \n \n %s" %
      [quest.title, quest.current_amount,
          quest.target_amount, quest.description]
  var label = quest_labels[quest_id]
  label.text = quest_data_text
```

With that, we close our quest system onboarding, and you are ready to understand how you will use it for our online multiplayer version of the prototype! You've seen it all, how objects can update quest progress, where quests are gathered and stored, how we load and save a player's progress in a given quest, how we implement a node to represent a quest in our game, and finally, how this all comes together to display the quest's information to the player in a UI element.

In the upcoming section, we will see how the World's Main node works. Its main responsibility is to ensure the game world is running as planned with all objects in their proper places.

Initializing the game world

To ensure that the game runs as we plan, at least its initialization, we have the Main node. It essentially spawns 30 Asteroid instances using the Asteroids node and creates an instance of the Player scene using the Players node. As explained at the beginning of the *Breaking down the World scene* section, the latter two nodes are spawners.

In this local gameplay prototype, the Main node is really simple, but keep it in mind, especially regarding its responsibility, when you start implementing the online multiplayer features. For reference, the Main node script is shown in the following code snippet:

```
extends Node

@onready var asteroid_spawner = $Asteroids
@onready var player_spawner = $Players
func _ready():
  for i in 30:
    asteroid_spawner.spawn()
  create_spaceship()

func create_spaceship():
  player_spawner.spawn()
```

Note that it has a create_spaceship() method instead of directly calling the player_spawner.spawn() method. This will help you with your job later on, so you can thank our fake team for making your job easier.

And with that, your onboarding process is done! We've seen how the player controls their spaceship, how the asteroids take hits and explode, increasing the player's progress in a quest, how the quest system works, and what it handles and outputs. We've also just seen how the game world initializes and establishes where each object should be and how many of them there should be.

Now, it's time for the magic. In the upcoming sections, we are going to see how we will turn this prototype into an online multiplayer game prototype where players can join anytime, so there won't be a lobby. We will also understand what we need to do to keep the players' world in sync with the server's world and how we separate the server and client's responsibilities using the same script. This will be useful especially when handling our databases to prevent players from cheating and completing quests without effort.

Logging the player in to the server

In this section, we will implement a different type of logging system. This time, the players don't have a lobby screen where they wait for other players to join a game and start a match. No, here the world is always active; it doesn't start only when the players ask the server to start the match or the game. And this type of connection requires a different approach.

The major problem is that since the game world is always active and running, the players who join this world need to sync their game instance to the server's game instance. This includes the position of objects, new objects that usually are not part of the world, for instance, other players and the number of objects (in our case, how many asteroids are currently available), and many other factors necessary to build a shared world.

It all starts with the player authentication, because now the server and the client are in different parts of the game life cycle; while the player is just opening the game, the server is already handling the game world.

Authenticating players

Don't panic, yet. Besides the conditions of authentication being different, the overall logic is very much the same as the one we've been using so far. The major difference here is that we will need to have an `Authentication` node dedicated to each side of the connection performing the authentication procedure according to the client or server's responsibilities.

These nodes will be on the two major points of interaction for each side of the connection:

- For the client, in this case the player, we will have the `Authentication` node on the `LoggingScreen` scene

- For the server we will have its `Authentication` node on the `World` scene itself, waiting for players to join

Note that for each side of this relationship, we will implement distinct authentication procedures. So, besides both nodes being called `Authentication` and having the same path, in other words, both being direct children of a parent node called `Main`, they will be totally different classes.

They will need shared methods, but we will see that the method implementations are different. This is all due to how RPCs work. Remember, when making an RPC, it will look for a node with the same node path in all peer game instances, and this node must have all the same methods as the one making the RPC, even if we are not calling these other methods. This means that the server side will share client-side methods, and vice versa.

This will get less confusing once we start to implement it, so let's open the `res://09.prototyping-space-adventure/LoggingScreen.tscn` scene and implement the client side of the authentication.

Implementing client-side authentication

In the `LoggingScreen` scene, you will notice a scene structure meant to be a simple logging screen where players insert their credentials, get authenticated, and log in to the game. This is a bit similar to the lobby we've been using in previous chapters, such as the one in *Chapter 8*, *Designing an Online Co-Op Platformer*.

This time, we don't have a panel showcasing the current players; this is not necessary as players can join the game and experience individually even when other players are not around. Note that the scene's script is attached to the `Authentication` node this time, instead of attached to the `Main` node.

This is because in the `World` scene, the `Main` node has other responsibilities, so it's better to delegate the authentication to an exclusive node. Due to that, `LoggingScreen`'s authentication was also delegated to its `Authentication` node.

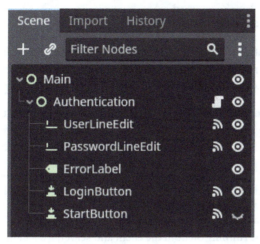

Figure 9.9 – The LoggingScreen scene's node hierarchy

Now, open the `res://09.prototyping-space-adventure/LoggingScreen.gd` script file. You will notice there are a lot of things in common with what we've created in *Chapter 3, Making a Lobby to Gather Players Together*, so let's focus on the necessary work we need to do this time:

1. Since this time we are only communicating with the server, we don't need to start the game on all peers, so in the `_on_StartButton_pressed()` callback, we need to send an RPC directly to the server asking it to start the game:

    ```
    func _on_StartButton_pressed():
      rpc_id(1, "start_game")
    ```

2. The server itself will then authenticate the player, and if everything goes well, the server will also call the client's `start_game()` method, which has a different implementation. In the client, `start_game()` is an RPC that only the network authority can call and is called locally. When called, it switches the current scene to the `next_scene` property, which in this case will be the `World` scene:

    ```
    @rpc("authority", "call_local")
    func start_game():
      get_tree().change_scene_to_file(next_scene)
    ```

Alright, most of the code is very similar to the one we have in the implementations of the Lobby scene. This one is cleaner as we've removed other methods such as the ones we used to display logged players or display avatars.

With these changes, this code is able to send a direct request authentication for the server and start the game on this instance of the game. In the next section, we will see the other side of this system, the server side.

Implementing server-side authentication

Now, the server-side authentication is a bit trickier. Previously, all that we needed to do was to handle the player's authentication requests. But now, since the authentication happens while the server is already running the game, we need to transfer the responsibility of setting up the hosting as well.

This means that if the current instance is the server, it will need to set up ENetMultiplayerPeer on top of authenticating players' credentials as well. Open the res://09.prototyping-space-adventure/Authentication.gd file and let's make the necessary changes.

Again, we will focus only on what we need to change based on the work we did previously, so feel free to go through the other parts of the script if you don't remember how this all works:

1. In the _ready() callback, let's create the ENet server using the default PORT. Remember, since this script will be running on both the client and server, we will need to check whether the current instance running is the server. For that, we use the multiplayer.is_server() method. After setting up the server, we load the users' database to properly authenticate them as usual:

    ```
    func _ready():
        if multiplayer.is_server():
            peer.create_server(PORT)
            multiplayer.multiplayer_peer = peer
            load_database()
    ```

2. The second thing we need to do is to connect as a client if the instance is not the server:

    ```
    else:
        peer.create_client(ADDRESS, PORT)
        multiplayer.multiplayer_peer = peer
    ```

3. The third thing we need to do is to set up the start_game() method to respond to all peers remotely, so this method won't be called locally in the server's instance. Inside this method, we will make an RPC to the peer that requested the game to start telling their instance to start the game.

This allows the server to dictate whether the player can or cannot join the game, and if the player happens to try to connect locally, simply pressing the start button won't actually start the game as their instance will hang, waiting for a response from the server:

```
@rpc("any_peer", "call_remote")
func start_game():
    var peer_id = multiplayer.get_remote_sender_id()
    rpc_id(peer_id, "start_game")
```

With these changes, we are able to have a server that can run an independent instance of the game and wait for players to join it. Using the `rpc_id()` method, we can pinpoint which peer we want to contact and establish direct communication between instances of the game. In this case, we did that with the server and the client, but we can do it between two clients as well if necessary.

In the next section, we are going to focus on how to sync the server's persistent `World` scene with the players' `World` scene, which may not reflect the current state of the shared game world.

Syncing the World scenes

When the player logs in to the game, their world will likely be different from the server's world. For instance, only the server should be able to spawn asteroids, and even if the client were able to spawn them, there would be nothing to guarantee they would be in the same position. So, this is the first thing we are going to fix.

In the next section, we will see how we can sync the `Asteroid` instances from the server's world into the client's world.

Syncing the asteroids

Open the `Asteroid` scene again and let's add a new `MultiplayerSynchronizer` as its child. This `MultiplayerSynchronizer` will replicate the asteroid's `position` and `top_level` properties. The following figure showcases the asteroid's **MultiplayerSynchronizer Replication Menu** settings.

Figure 9.10 – The Asteroid's MultiplayerSynchronizer Replication Menu settings

Then, we are going to use something really interesting regarding network replication. The `MultiplayerSynchronizer` node has a property called **Public Visibility**, which essentially enables the replication of these properties to all connected peers. But we don't want that. We only want it to replicate these properties once the players are logged in and within the `World` scene. So, we will toggle that property off. In the following screenshot, we can see what this property should look like in the **Inspector**.

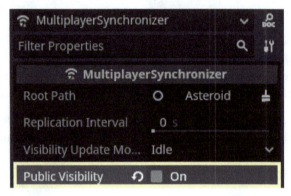

Figure 9.11 – The asteroid's MultiplayerSynchronizer Public Visibility property in the Inspector

Before we move on to the next necessary steps, I want to present you with a trick that will help you synchronize relevant objects all at once. Add `MultiplayerSynchronizer` node inside a group so that you can perform a group call using `SceneTree` later on. In this case, to be clearer about the group's intent, let's call it `Sync`. The following screenshot showcases this group with the `Asteroid` node's `MultiplayerSynchronizer` node inside of it.

Figure 9.12 – The asteroid's MultiplayerSynchronizer inside the Sync group

With that, the `Asteroid` instances are ready to replicate their properties on the client's `World` instance. But there's still a problem. The client's `World` instance should not create these `Asteroid` instances; instead, only the server should be able to spawn these objects. So, let's open the `World` scene and set it up for synchronization:

1. The first thing we need to do is to add a `MultiplayerSpawner` node that points to the `Asteroids` node. Let's call it `AsteroidsMultiplayerSpawner`, and it should have the `Asteroid` scene set up as its first and only element in the **Auto Spawn List** property. We can see these properties configured in the following screenshot:

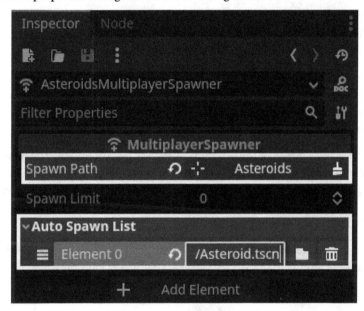

Figure 9.13 – The world's AsteroidsMultiplayerSpawner properties in the Inspector

This will ensure that all the instances of the server's world asteroids will also exist in the client's world as well. But note that until now, they will only be spawned, but not synced yet. So, let's fix that. Open the `World` script and let's set it up for the sync logic.

2. First things first, in the `_ready()` callback, we need to prevent the `World` node from spawning asteroids if it isn't the server. It should request synchronization from the server instead. For that, it will make an RPC to the server's `sync_world` method, which we will create in the next step:

```
func _ready():
  if not multiplayer.get_unique_id() == 1:
    rpc_id(1, "sync_world")
  else:
    for i in 30:
      asteroid_spawner.spawn()
```

3. Then, let's create the `sync_world()` RPC method, which can be called by any peer locally. It needs to be called locally because we will tell the server's `Asteroid` instances' `MultiplayerSynchronizer` nodes, which are in the `Sync` group, to add the player to their visibility list, effectively syncing the `Asteroid` instances.

```
@rpc("any_peer", "call_local")
func sync_world():
    var player_id = multiplayer.get_remote_sender_id()
    get_tree().call_group("Sync", "set_visibility_for
        ",player_id, true)
```

`set_visibility_for()` is a method from the `MultiplayerSynchronizer` node that adds a peer to its visibility list, which basically means a whitelist of peers it should synchronize to.

For that, it uses the peer's ID and receives a Boolean to tell it whether this peer should or shouldn't see the replication of the properties set in **MultiplayerSynchronizer Replication Menu**. This works because we get the ID of the player who requested to sync using the `multiplayer.get_remote_sender_id()` method, so whoever requests to sync will be synced.

This is all we need for the syncing of the asteroids. Now, we are still missing the players and their spaceships, right? In the next section, we will see how to remotely create `Player` instances on all connected peers, sync their spaceships, and only allow their owner to control the spaceship.

Syncing the players

It's time to put our players together in this vast world. In this section, we will understand how we sync players' scenes that were already in the World when another player joins the game.

Let's get started:

1. Still in the `World` script, we are going to turn the `create_spaceship()` method into an RPC method that any peer can call remotely:

```
@rpc("any_peer", "call_remote")
func create_spaceship():
```

2. Due to how the `Players` spawner works, we won't be able to do the proper renaming and identification of the newly created spaceship before it syncs to other peers. So, the `create_spaceship()` method takes the responsibility of spawning spaceships as well.

    Before we add the `spaceship` instance as a child of the `Players` node, we will set its name to match the player's peer ID. This ensures this instance has a unique name and we can use this name to identify the proper authority of the instance:

```
    var player_id = multiplayer.get_remote_sender_id()
    var spaceship = preload("res://09.prototyping-space-
    adventure/Actors/Player/Player2D.tscn").instantiate()
```

```
spaceship.name = str(player_id)
$Players.add_child(spaceship)
```

3. Now we come to a very important part. We are going to implement the setup_multiplayer() method in the player, which essentially does the same thing as the one we made in *Chapter 8, Designing an Online Co-Op Platformer*. So, we can make an RPC to this function here after waiting for 0.1 seconds:

```
await(get_tree().create_timer(0.1).timeout)
spaceship.rpc("setup_multiplayer", player_id)
```

With that, whenever a player asks the server's World instance to create a spaceship, it will instantiate a Player scene, assign it a unique ID, and ask it to configure its multiplayer settings. Remember, since we are doing this using an RPC, this means this Player instance will configure its multiplayer settings in all currently connected peers. But as it is now, only the server has an instance of this Player node.

4. To fix that, we are going to add another MultiplayerSpawner node called PlayersMultiplayerSpawner to the World scene. Its **Spawn Path** should point to the Players node and its **Auto Spawn List** should have the path to the Player scene, res://09.prototyping-space-adventure/Actors/Player/Player2D.tscn. In the following screenshot, we can see these properties set up in the **Inspector**.

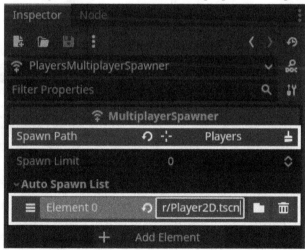

Figure 9.14 – The World scene's PlayersMultiplayerSpawner node properties in the Inspector

Now, since the instances created by the PlayersMultiplayerSpawner node will still not be configured yet, we also need to call their setup_multiplayer() method as soon as they spawn.

5. For that, let's connect the `PlayersMultiplayerSpawner`'s spawned signal to the `World` scene's `Main` node script, and inside then `_on_players_multiplayer_spawner()` callback, we make an RPC on the recently spawned node passing `set_up_multiplayer` as argument. This time, we will use the node's name as an argument instead of `player_id`. This is because we don't have access to the ID of the player that is supposed to own this instance, so we can use the instance name instead:

    ```
    func _on_players_multiplayer_spawner_spawned(Node):
        Node.rpc("setup_multiplayer", int(str(Node.name)))
    ```

6. With that, every time a player joins the game, the server will create a new `Player` instance for them and set this instance up. This also works for `Player` instances that are already in the server's `World`.

    If a player joins the game, the server will spawn all other `Player` instances that are also currently playing. Now, we need the `Player` scene itself to sync its relevant properties to its peers and to implement its `setup_multiplayer` method.

7. Open the `res://08.designing-online-platformer/Actors/Player/Player2D.tscn` scene, and let's start by adding `MultiplayerSynchronizer` node as a child of the `Player` node. This `MultiplayerSynchronizer` node should sync the `Player` instance's `position` and `top_level` properties and the `Spaceship` node's `position` and `rotation` properties. The following screenshot showcases the `Player` scene's **MultiplayerSynchronizer Replication Menu:**

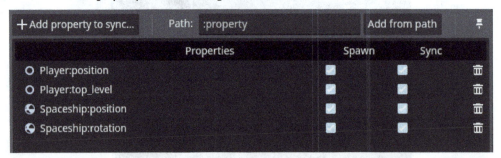

Figure 9.15 – The Player's MultiplayerSynchronizer Replication Menu

Since this `MultiplayerSynchronizer` node will work by syncing physics properties, we need to set its **Visibility Update Mode** property to **Physics**. This will prevent some weird behaviors such as bodies overlapping and unhandled or mishandled collisions.

Now, let's implement the `setup_multiplayer()` method. In the following instructions, we are going to create an RPC method that checks whether the current instance is a competence of the current player and disables some important processes to prevent interaction from players into instances they don't own, as well as preventing overwriting properties synchronized by the network.

8. In the `Player.gd` script, let's start by creating an RPC method called `setup_multiplayer` that any peer can call locally. It should receive `player_id` as an argument:

```
@rpc("any_peer", "call_local")
func setup_multiplayer(player_id):
```

9. Then, inside this method, we need to compare whether the `player_id` value received as an argument matches the current player's peer ID. We will store this information in a variable called `is_player` to use for further reference:

```
var self_id = multiplayer.get_unique_id()
var is_player = self_id == player_id
```

10. With this information in hand, we can set up the player's processes. We also need to disable `camera` if this isn't the current player and make `camera` the current camera if it is:

```
set_process(is_player)
set_physics_process(is_player)
camera.enabled = is_player
if is_player:
  camera.make_current()
```

11. Finally, we set the player's multiplayer authority to be `player_id`. This will ultimately prevent this client from making any changes on this `Player` instance and propagating them on the network:

```
set_multiplayer_authority(player_id)
```

With that, we have finished all we need to do in order for players to spawn and control their spaceships in the game. If you test the game now, you will be able to put two players together and control their spaceships only on their game instances. For reference, the following code snippet showcases the complete `setup_multiplayer()` method's code implementation:

```
@rpc("any_peer", "call_local")
func setup_multiplayer(player_id):
  var self_id = multiplayer.get_unique_id()
  var is_player = self_id == player_id
  set_process(is_player)
  set_physics_process(is_player)
  camera.enabled = is_player
  if is_player:
    camera.make_current()
  set_multiplayer_authority(player_id)
```

Throughout this section, we learned how we can spawn already existing objects to a player's game instances when they log in when the game world is already running. We also saw how we can selectively synchronize objects using the `MultiplayerSynchronizer.set_visibility_for()` method.

On top of that, we used the `MultiplayerSpawner.spawned` signal to configure spawned instances of an object on the client's side. In our case, we needed to configure the player multiplayer settings. To do that, we created a method that checks whether this instance belongs to the current player, properly disabling or enabling its processing and camera and setting its multiplayer authority accordingly.

In the next section, we are going to learn how we can separate some responsibilities in the game to prevent cheating and to establish a more coherent multiplayer system, for instance, preventing a player's game instances from deleting asteroids as they are the responsibility of the server.

We will also see how we can sync players' actions on all peers. This will be useful to replicate actions locally; for instance, when a player shoots on their game instance, all other game instances should perform the shooting as well.

Separating server and client responsibilities

Now that we have players sharing the same world, we need to establish which actions they are responsible for and which actions are part of the server's responsibility. For instance, if a player shoots on their game instance and their bullet damages an asteroid but this asteroid was already destroyed by another player, what should happen? For this kind of situation, the server is the perfect mediator to prevent instance conflicts.

With all this context in place, players tell all peers, including the server, to update their `Player` instance according to their actions, but only the server should have the authority to manage the actual impact of these actions in the game world, such as if the player managed to destroy an asteroid or not. In the next section, we are going to understand how players can sync their actions, not only their objects' properties, across all network-connected peers.

Shooting bullets on all instances

Since the `Player` scene's `Spaceship` node movement is already synced by the `MultiplayerSynchronizer` node, we can focus our efforts on syncing the `Bullet` instances' movement now. One way we could do that would be to use the `MultiplayerSpawner` and `MultiplayerSynchronizer` nodes to spawn `Bullet` instances remotely and replicate their position, rotation, and so on. But instead, we can make an RPC telling all `Spaceship` instances to call the `fire()` method on their `Weapon2D` nodes.

This is a quick and cheap way to do this. As the `Bullet` nodes have a constant trajectory, there's no reason to sync their movement properties. The only thing relevant is where they start, that is, the spawning position, and the direction they should move. Both these properties are being synced already by the `Player` node's `MultiplayerSynchronizer` node. So, we can leverage them. Nice trick, right?

To do this, open the `res://09.prototyping-space-adventure/Actors/Player/Player2D.gd` script, and in the `_process()` callback, change the `weapon.fire()` line to `weapon.rpc("fire")`, as shown in the following code snippet:

```
func _process(delta):
  if Input.is_action_pressed(shoot_action):
    weapon.rpc("fire")
```

If you test the game now, you should see the `Player` instances' `Spaceship` nodes shooting bullets on all game instances: server and clients. In the following screenshot, we can see a player shooting being replicated on the server's game instance:

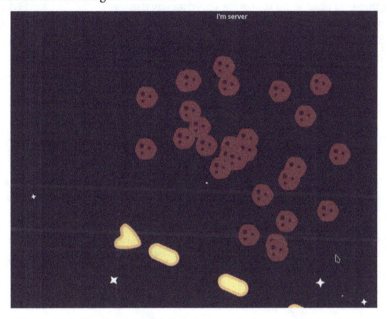

Figure 9.16 – The Payer scene's Spaceship node shooting Bullet instances
on the server's game instance using an RPC function

Now that the `Player` instances can shoot bullets on every peer instance of the game, we need to understand who should manage the damage calculation and the destruction of the objects the `Bullet` nodes hit. We've already talked about that, and this responsibility is the server's.

But how do we do that? In the next section, we are going to break down the `Asteroid` nodes' behavior in order to separate what should happen on the client's side and what should happen on the server's side.

Calculating an asteroid's damage

Here, it's time to use the `multiplayer.is_server()` method extensively. We need to break down the `Asteroid` node's behavior and establish what should happen when the `Bullet` instances on a client's game instance hit an `Asteroid` node and what should happen when these `Bullet` instances hit an `Asteroid` node on the server side.

Open the `res://09.prototyping-space-adventure/Objects/Asteroid/Asteroid.gd` script, and let's implement the damage-taking behavior, respecting the responsibilities of each side of the connection:

1. The first thing we need to do is prevent applying any damage to the `Asteroid` node if the current peer isn't the server. So, in a client's game instance, the `Bullet` node should hit the `Asteroid` node and disappear, but should not apply damage to the `Asteroid` node. To do that, in the `_on_hurt_area_2d_damage_taken()` callback, we are going to check whether the peer is the server, and if it is, we call the `apply_damage()` method, passing damage as an argument:

```
func _on_hurt_area_2d_damage_taken(damage):
    if multiplayer.is_server():
        apply_damage(damage)
```

2. Now, in the `apply_damage()` method, after doing the proper calculation, the server must tell the `Asteroid` instances on all peers to play their proper animation, either playing `"hit"` or `"explode"`. But here's the trick: the `animator` doesn't have an RPC method to do that.

So, instead, we are going to extract this behavior into two RPC methods and call these methods. These RPC methods should be called only by the network authority, and they should also be called locally:

```
func apply_damage(damage):
    health -= damage
    if health < 1:
        rpc("explode")
    elif health > 0:
        rpc("hit")

@rpc("authority", "call_local")
func explode():
    animator.play("explode")

@rpc("authority", "call_local")
func hit():
    animator.play("hit")
```

3. The last thing we need to do is to allow only the server to call the queue_free() method on the Asteroid nodes, preventing players from cheating and completing the quest in unpredictable ways. To do that, in the _on_animation_player_animation_finished() callback, we are going to check whether the peer is the server and call queue_free() if the current animation is "explode":

```
func _on_animation_player_animation_finished(anim_name):
  if multiplayer.is_server():
    if anim_name == "explode":
      queue_free()
```

Since the server, which is the Asteroid node's multiplayer authority, is removing the Asteroid instance from its SceneTree, the World node's AsteroidsMultiplayerSpawner node will ensure that the Asteroid instances spawned on the clients' game instances will also be removed as well. Isn't the Godot Engine's Network API clever?

With that, each side of the connection is performing its duty. The client side plays animations based on the Asteroid node's state, while the server side deals with the actual impacts of Bullet nodes hitting Asteroid nodes. In this section, we saw how we can work around the issue when a behavior needs to be replicated remotely on all peers, but the built-in class, for instance the AnimationPlayer, doesn't have a way to do it.

We also learned how to separate things and give each side of the connection the power to execute their responsibility. While the client side must instantiate bullet's and do all the processing of the shooting, the server side does its part by processing the damage dealt by the Bullet nodes and handling the Asteroid node's life cycle.

In the next section, we are going to strengthen this knowledge by applying the same principles to the quest system. How does the player retrieve their quests? What is the client's responsibility? Should the client store the player's progress? What about the server? How does it handle clients' requests and maintain information consistently between play sessions? That's what we are going to talk about.

Storing and retrieving data on the server

It's time to handle a sensitive topic when it comes to online adventure games: databases. Note that in this book, we are not focusing on the best and most secure way to handle and protect a database. Instead, we are practicing and understanding what the Godot Engine Network API allows us to achieve and exploring its possibilities.

That said, in this section, we are going to implement the necessary steps to establish a communication channel where the client can retrieve its quests data from the server and send their progress updates to the server.

To do that, we are going to work with the two main classes in our quest system, `QuestSingleton` node and the `QuestDatabase` node. But before we set these classes up for this new challenge, we need to change how the database is structured. Since now the `QuestDatabase` node will work by delivering and handling multiple players' data, the `PlayerProgress.json` file needs to have its data linked to a user. So, let's create these fake users, matching the ones in `FakeDatabase.json` file, and store this arbitrary data.

Open the `res://09.prototyping-space-adventure/Quests/PlayerProgress.json` file and create two new keys, one for `user1` and another for `user2`. Then, add some manual data matching the original structure for each user key:

```
{
  "user1":{
    "asteroid_1":{
      "completed":false,
      "progress":4
    }
  },
  "user2":{
    "asteroid_1":{
      "completed":false,
      "progress":2
    }
  }
}
```

Now, with this in mind, remember that we can, and should, use the user we have stored in the `AuthenticationCredentials` singleton to refer to any sensitive user data. This is important because it's how we are going to properly manage users' requests and deliver the data accordingly. In the next section, we are going to see how `QuestSingleton` node should retrieve and update quest data.

Implementing the quest system's client side

It's time to finally have our quest system ready to work in a network where players can retrieve quests from a remote database, make progress on them, and store their information for the future. In this section, we are going to see what the `QuestSingleton` node's role is in this online multiplayer environment.

So, let's open the res://09.prototyping-space-adventure/Quests/QuestSingleton.
gd script and get started with that. In the following instructions, we will see how we can retrieve and
update quest data in a remote QuestDatabase:

1. The retrieve_quests() method needs a complete, but simpler and more elegant, revamp.
    The easiest way to create new quests in this online environment is to make the client request
    the remote QuestDatabase to create them remotely. We will see how this happens in depth
    in the *Implementing the quest system's server* section.

2. But for now, we will wait for 0.1 seconds to guarantee that everything is in place, and
    then we can make an RPC on the QuestDatabase node's get_player_quests()
    method if this QuestSingleton node is not the server's. Remember to pass
    AuthenticationCredentials.user property as an argument to the get_player_
    quests() method:

    ```
    func retrieve_quests():
      if multiplayer.is_server():
        return
      await(get_tree().create_timer(0.1).timeout)
      QuestDatabase.rpc_id(1, "get_player_quests",
        AuthenticationCredentials.user)
    ```

3. Now, since the server's QuestDatabase node will create the quests on the client's
    QuestSingleton node, we need to turn the create_quest() method into an RPC
    method that only the authority can call remotely:

    ```
    @rpc("authority", "call_remote")
    func create_quest(quest_data):
    ```

4. Finally, in the increase_quest_progress() method, we need to call the
    QuestDatabase.update_player_progress() method using an RPC directly to
    the server as well. Let's not forget to also pass AuthenticationCredentials.user
    as an argument as well:

    ```
    func increase_quest_progress(quest_id, amount):
      if not quest_id in quests.keys():
        return
      var quest = quests[quest_id]
      quest.current_amount += amount
      QuestDatabase.rpc_id(1, "update_player_progress",
          quest_id, quest.current_amount, quest.completed,
            AuthenticationCredentials.user)
    ```

With that, the `QuestSingleton` node will ask the server's `QuestDatabase` node to create the current player's quests, and whenever the player makes progress in a quest, the client's `QuestSingleton` node updates the server's `QuestDatabase` node about the current quest's progress, ultimately delegating to the server the responsibility of handling this important data.

Now, my friends, comes the last piece of our puzzle. In the next section, we are going to understand how the server side of this system works now that it is put in an online multiplayer environment.

Implementing the quest system's server side

In this section, we will add the necessary functionalities to the `QuestDatabase` node in order for it to properly provide and store players' quest data. To do that, we are going to use RPCs extensively together with the `multiplayer.is_server()` method to prevent some behaviors from happening on the players' instances of the `QuestDatabase` node.

We do this mainly to maintain the quest data only on the server side, so clients don't run into the temptation to cheat directly on their machines. This will be an extensive section, so take a breather. We've been through quite a bit of information up to this point. Take a pause, and once you feel ready, come right back.

Ready? Okay, open the `res://09.prototyping-space-adventure/Quests/QuestDatabase.gd` script and let's get started. In the upcoming instructions, we are going to tweak the current `QuestDatabase` node's methods and even create new ones in order to make it work as it should, starting with…the `_ready()` method:

1. In the `_ready()` callback, we should only load the database files if the current peer is the server:

    ```
    func _ready():
      if multiplayer.is_server():
      load_database()
    ```

2. We do the same thing in the `_notification()` callback. We should only store the files if the current peer is the server, so let's add this check together with the notification check:

    ```
    func _notification(notification):
        if notification == NOTIFICATION_WM_CLOSE_REQUEST
          and multiplayer.is_server():
          store_database()
    ```

3. Moving on to `get_player_quests()`, it should now be an RPC method that any peer can call remotely:

    ```
    @rpc("any_peer", "call_remote")
    func get_player_quests():
    ```

4. Since it's now an RPC, we will store the peer ID of the client that requested this data so we can use this later on when we respond to the request:

```
func get_player_quests():
    var requester_id = multiplayer.get_remote_
        sender_id()
```

5. Then, add an argument called `user` to the method's signature so we know which key to look at in the `progress_database` dictionary. Remember, now each user has their own key with their progress data in the `PlayerProgress.json` file, so this is how we access the proper user data:

```
func get_player_quests(user):
    var requester_id = multiplayer.get_remote_
        sender_id()
    var quests = {}
    for quest in progress_database[user]:
```

6. Now, we have some changes to make in the `get_progress()` and `get_completion()` method signatures, right? We need to add the user as an argument. So, let's call them using this argument:

```
for quest in progress_database[user]:
    var quest_data = {}
    quest_data["id"] = quest
    quest_data["title"] = get_title(quest)
    quest_data["description"] = get_description(quest)
    quest_data["target_amount"] = get_target_amount
        (quest)
    quest_data["current_amount"] = get_progress
        (quest, user)
    quest_data["completed"] = get_completion
        (quest, user)
    quests[quest] = quest_data
```

7. Now, we need to fix the methods' signatures to match the previous changes and allow them to receive a `user` argument and access the `user` key in the `progress_database`:

```
func get_progress(quest_id, user):
    return progress_database[user][quest_id]["progress"]

func get_completion(quest_id, user):
    return progress_database[user][quest_id]
        ["completed"]
```

8. Back to the `get_player_quests()` method – for each `quest` key found in `progress_ database` dictionary for a user, we will make an RPC directly to the `create_quest()` method of the client's `QuestSingleton` node, passing the `quest_data` dictionary we just created:

```
for quest in progress_database[user]:
    var quest_data = {}
    quest_data["id"] = quest
    quest_data["title"] = get_title(quest)
    quest_data["description"] = get_description
        (quest)
    quest_data["target_amount"] = get_target_
        amount(quest)
    quest_data["current_amount"] = get_progress
        (quest, user)
    quest_data["completed"] = get_completion
        (quest, user)
    quests[quest] = quest_data
    Quests.rpc_id(requester_id, "create_quest",
        quest_data)
```

9. Finally, let's also turn `update_player_progress()` into an RPC method that any peer can call remotely. It should also receive a `user` argument to update the progress on the proper user key in `progress_database`. This should only happen if this is the server's instance of the `QuestDatabase` node, of course:

```
@rpc("any_peer", "call_remote")
func update_player_progress(quest_id, current_amount,
    completed, user):
  if multiplayer.is_server():
    progress_database[user][quest_id]["progress"] =
        current_amount
    progress_database[user][quest_id]["completed"] =
        completed
```

And that wraps up our quest system and database management logic on both the client and server sides. If you test the `res://09.prototyping-space-adventure/MainMenu.tscn` scene and log in as a user, you will be able to see `QuestPanel` displaying the quest data correctly with the current player's progress on the quest. In the following screenshot, we can see `user2`'s quest information:

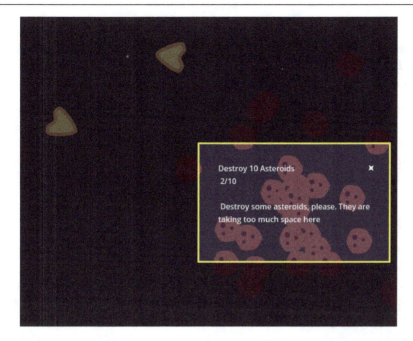

Figure 9.17 – QuestPanel displaying user2's Destroy 10 Asteroids quest information

The following screenshot displays user1's quest information. So, we can assume that our system is working as it should, properly loading, displaying, and modifying players' quest progress:

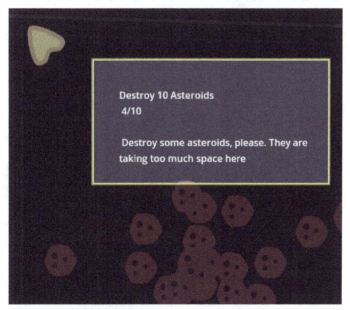

Figure 9.18 – QuestPanel displaying user1's Destroy 10 Asteroids quest information

With that, our top-down space adventure prototype is ready! Congratulations on making it all the way to this point. This chapter tested out everything we've seen so far extensively, and it closes *Part 2, Creating Online Multiplayer Mechanics*, of this book, where we created five prototypes to learn the ins and outs of Godot Engine's high-level Network API.

Hopefully, by now, you feel confident in building and testing more prototypes on your own. In *Part 3, Optimizing the Online Experience*, which we are going to start in the next chapter, we are going to see how we can use the available tools to improve upon the experiences we created in *Part 2*. We will talk about how to debug and profile the network, optimize data requests, implement optimization techniques such as prediction and interpolation, cache data, and more.

Summary

In this chapter, we learned how to allow players to join in the middle of a game run; how to synchronize their game instances; how to load, retrieve, send, and store information on a remote database; how to create a quest system; and overall, how to structure the very basics of an online multiplayer adventure game. In the next chapter, we are going to learn how to debug and profile the network so we can find bottlenecks and potential areas of improvement and optimization for our games. See you there!

Part 3:
Optimizing the Online Experience

One of the core aspects of developing an application, especially a game, is to make the experience smooth and keep it without any hiccups or lag. When talking about games, which are real-time experiences, this is even more relevant. So, throughout this part of the book, we learn about the debugging and profiling tools necessary to assess potential bottlenecks and then we implement techniques to effectively optimize the network usage of the final project from *Part 2*.

This part contains the following chapters:

- *Chapter 10, Debugging and Profiling the Network*
- *Chapter 11, Optimizing Data Requests*
- *Chapter 12, Lag and Packet Loss Compensation*
- *Chapter 13, Caching Data to Decrease Bandwidth*

10

Debugging and Profiling the Network

With *Chapter 9, Creating an Online Adventure Prototype*, we concluded *Part 2, Creating Online Multiplayer Mechanics*, of our journey, where we learned how we can use Godot Engine's High-Level Network API to turn local gameplay mechanics into online multiplayer mechanics. Now, it's time to go beyond implementation and start the optimization of our mechanics. This chapter inaugurates *Part 3, Optimizing the Online Experience*, of our journey through creating online multiplayer games with Godot Engine.

It's important that you have read, understood, and implemented the content provided in *Chapter 9, Creating an Online Adventure Prototype,* because we are going to use the final project as our main subject through the following chapters in *Part 3*.

In this specific chapter, we are going to understand how we can use Godot Engine's built-in **Debugger** dock to assess and profile our game performance. For that, we are going to understand and use tools such as the **Profiler** tab, which helps us understand the rendering time of each frame by showing how many resources each function takes and how this impacts the processing time. We are also going to see one of the most important tools for our network craft, the **Network Profiler** tab, which helps us understand how many network resources each node's **Remote Procedure Call** (RPC) functions are taking and how our `MultiplayerSynchronizers` are performing, giving us a good overview of potential issues in our network implementation. Finally, we are going to learn how to use the Debugger dock's **Monitors** tab and the `Performance` singleton to figure out potential bottlenecks in our game and gather data to design potential solutions.

By the end of this chapter, you will understand how to use the powerful Debugger tools, and the elements in the following figure won't scare you anymore; instead, they will be some of your most reliable allies:

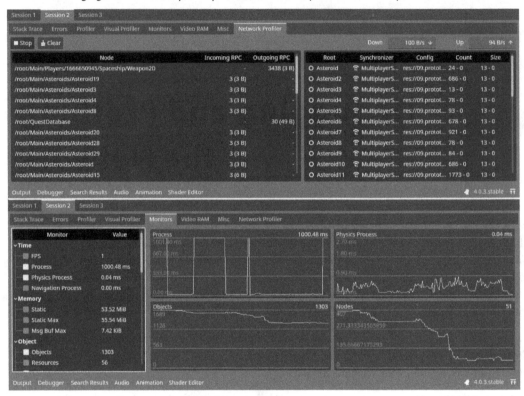

Figure 10.1 – The Debugger's Network Profiler (top) and Monitors
(bottom) showing and plotting profiling data

Don't be surprised if you come back to this figure at the end of the chapter and understand what each of those graphs and charts means. You will get used to them, as they will appear in abundance throughout the next chapters, especially *Chapter 11, Optimizing Data Requests*.

Technical requirements

As mentioned previously, it's crucial that you've read and followed the instructions provided in *Chapter 9, Creating an Online Adventure Prototype*. Here, in this chapter, we are going to use the final product you should have by the end of the previous chapter. You can access the resources for this chapter in the repository provided in the following link:

```
https://github.com/PacktPublishing/The-Essential-Guide-to-Creating-
Multiplayer-Games-with-Godot-4.0
```

With the result of *Chapter 9, Creating an Online Adventure Prototype,* ready, we can move on to understanding how we can improve it.

Introducing Godot's Debugger

The Debugger is a developer's best friend. Most of the work we do doesn't have anything to do with creating and implementing features; instead, it has everything to do with assessing potential problems these implementations cause and fixing them. The **Debugger** dock is where Godot Engine talks to us, showing errors, warnings, resource consumption, object count, and more. So, we should listen carefully and properly address the issues and data it shows us. We can even ask it to track custom data, as we are going to see in the *Using the Monitors tab* section.

If you have been developing games with Godot Engine for enough time to run into errors, you have probably stumbled on the **Debugger** dock more than you'd like to, right? In this section, we will go in-depth to understand how to turn it into our best friend and actually wish it pops up. Let's start by understanding each of its tabs, how to read them, and what to expect from them, starting with the most common and probably the one you've already had a hard time with: the **Stack Trace** tab.

Mastering the Stack Trace tab

When you click on the **Debugger** dock, Godot Engine's editor will open the **Stack Trace** tab. Let's use the following figure to navigate it and understand what each of its elements does.

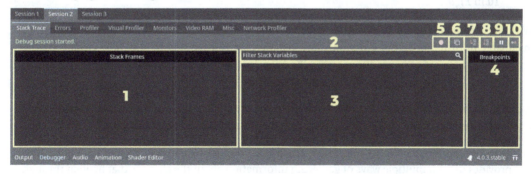

Figure 10.2 – The Debugger dock's Stack Trace tab and its elements

You can see that each element of the **Stack Trace** tab in the figure is associated with a number, which will facilitate a better understanding. In the following list, we have the element's name and a brief explanation about it:

- The **Stack Frames** panel is the stack of functions that leads to an error or a breakpoint (highlighted and marked with **1** in *Figure 10.2*).

- The **Filter Stack Variables** field is where you can filter variable names to display them in the panel below (highlighted and marked with **2** in *Figure 10.2*).

- The **Members** panel is where you can find the variables within a given script including temporary variables and scope-specific variables. Here, you can also see and edit their values (highlighted and marked with **3** in *Figure 10.2*).

- The **Breakpoints** panel is where you can see information about a breakpoint reached in the script of a given instance (highlighted and marked with **4** in *Figure 10.2*).

- The **Skip Breakpoints** button, when toggled on, allows the execution of the game to run ignoring breakpoints (highlighted and marked with **5** in *Figure 10.2*).

- The **Copy Error** button copies the current error, if any, to your clipboard (highlighted and marked with **6** in *Figure 10.2*).

- Clicking on the **Step Into** button when the application is paused, including when it reaches a breakpoint, will execute the next script instruction (i.e., line). It will enter in indented blocks it would naturally go into, executing the whole code (highlighted and marked with **7** in *Figure 10.2*).

- Clicking on the **Step Over** button when the application is paused, including when it reaches a breakpoint, will execute the next script instruction (i.e., line) but skip indented blocks (highlighted and marked with **8** in *Figure 10.2*).

- The **Break** button pauses the application as if it reached a breakpoint (highlighted and marked with **9** in *Figure 10.2*).

- The **Continue** button resumes the application if it was paused (highlighted and marked with **10** in *Figure 10.2*).

With these elements at our disposal, we have the ability to experiment with our scripts and gather invaluable information about our game. For instance, we can see step by step how Godot Engine processes a given set of instructions using the **Step Into** button and see the stack of functions it executes and how the objects' variables change with each step.

A cool tip to use this to its fullest is to not be afraid to add breakpoints all over your scripts to understand when, what, how, and why your objects change and the whole chain of events that caused such changes.

In this section, we've gone through the **Stack Trace** tab, which gives us an overview of our game's flow and provides us with multiple ways of gathering information about the changes that happen through this flow, allowing us to understand the whole chain of cause-effects that led to a given change. This is especially helpful together with our next tab, the **Errors** tab. Let's talk about it in the next section.

Debugging with the Errors tab

It may sound weird but, in many situations, you might wish for Godot Engine to prompt an error, especially when dealing with network features, as sometimes you are left waiting for something to happen. And if the packets sent don't reach their destination, you will be hanging there waiting for an error to pop up, but packets not reaching their destination isn't an error in itself. Still, it's an undesired situation that can leave you confused.

The **Errors** tab is where you work with thousands of other developers, who worked on the development of Godot Engine's core and identified thousands of errors and documented them so that when they happen, you have some light on the issue and are able to fix it.

However, not only errors are displayed in this tab. The **Errors** tab also shows warnings about your script. They don't necessarily break your application but are something you should be aware of and make a decision on. For instance, it's common to get warnings about arguments in a function that are not being used in the function's implementation. The following figure displays the **Error** tab and its elements associated with numbers, just like in the *Mastering the Stack Trace tab* section:

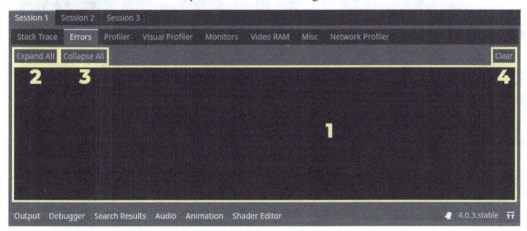

Figure 10.3 – The Session 1 Debugger dock's Errors tab and its elements

Now, let's understand what each of these elements is and how they can be useful for us:

- The **Errors and Warnings** panel is where all the warnings and fatal and non-fatal errors are displayed. You can click on an error or warning to expand it and go to the script line that triggered it. You can also double-click to expand an error or warning and display the code stack that led to the error. When you double-click an expanded error or warning, you collapse it (highlighted and marked with **1** in *Figure 10.3*).

- The **Expand All** button expands all the errors and warnings (highlighted and marked with **2** in *Figure 10.3*).

- The **Collapse All** button collapses all the errors and warnings (highlighted and marked with **3** in *Figure 10.3*).

- The **Clear** button empties the **Errors and Warnings** panel (highlighted and marked with **4** in *Figure 10.3*).

Something interesting about dealing with errors and warnings is that you can create your own error or warning messages. This is especially good when working with your teammates, but also, since we are working with more than one instance of the game running, it's a good way to compartmentalize messages to their instance's **Error** tab. As the **Output** dock is shared between all instances, it may get really confusing to identify where a `print()` statement is coming from. So you can use the `push_error()` and `push_warning()` built-in methods instead and Godot will only show them in the game session's **Debugger** dock that triggered the error or warning. The following figure showcases the **Error** tab of **Session 3** with a custom warning expanded so we can see where it comes from:

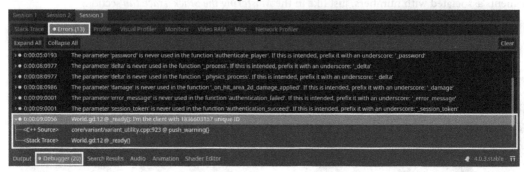

Figure 10.4 – The Session 3 Errors tab highlighting a custom warning among other built-in warnings

Notice that the **Debugger** button at the bottom tells us there's a total of 20 errors and warnings, but when we open the **Errors** tab in **Session 3**, there are only 13. This is because the other errors come from other sessions and are in their respective **Errors** tabs.

With this powerful tool in our arsenal, we can trigger all sorts of errors and warnings in each individual game session, allowing us to distinguish which session is the server, which ones are players if any peer is getting a specific error that others aren't, and so on. In the next section, we will talk about our first performance-based debugging tab, the **Profiler** tab, where we can see how our game is performing, how many resources it is taking, and which objects and functions are taking the most out of our computer.

Exploring the Profiler tab

Most developers are always looking for the most efficient, cheap, and ingenious optimizations to make their code run in a toaster. Well, while this is a nice and beautiful fantasy, the reality is that you shouldn't be so focused on optimizing your code unless you really need it. There's a saying in the industry that states, "*Premature optimization is the doom of an application.*"

Focus on the *premature* word here.

So, if premature optimization is something bad, but optimization in itself is something good, when is the right time to optimize your game or application? The answer is not set in stone and there's no clear point that we can just point out and say "*Here, after X days of development, it's time to optimize,*" or "*After you reach 80% of production, it's the sign to optimize.*" No, instead, you should address issues as they show up and create the habit of diagnosing your game's performance and deciding whether, based on your audience's computers' specs, you are going to need to squeeze some resources or not. This can happen on production day 1, or years after launching the game.

So, you need to engage in the habit of looking at how your game is performing and looking for areas of improvement regularly.

The **Profiler** tab is one of your best allies in optimization. It is in this tab that you will see rendering time, physics simulation time, audio processing time, and even how much time each of your custom script functions is taking to process and how many times they were called. Let's take a look at the following figure and understand how the **Profiler** tab displays all this information.

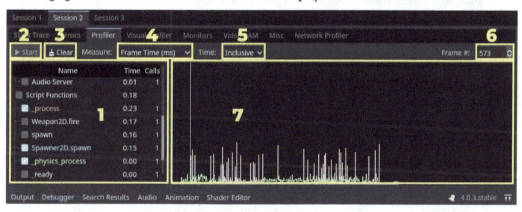

Figure 10.5 – Session 2's Debugger dock Profiler tab and its elements

Let's understand the role of each of these elements, again, following their number in the figure:

- The **Functions** panel displays the currently available functions that the profiler can track (highlighted and marked with **1** in *Figure 10.5*).

- The **Start** button initializes the measurement. Note that without toggling this on, the profiler won't do anything. Profiling is quite resource-intensive, so by default, it's off (highlighted and marked with **2** in *Figure 10.5*).

- The **Clear** button clears the current data gathered and displayed (highlighted and marked with **3** in *Figure 10.5*).

- The **Measure** drop-down menu allows us to change the type of data we want to measure (highlighted and marked with **4** in *Figure 10.5*). The current options are as follows:

  - **Frame Time (ms)** is how many milliseconds Godot Engine takes to process a frame.

  - **Average Time (ms)** is how long a function takes to process. This averages the time of each call of any given function.

  - **Frame %** is the percentage a given function takes to process relative to the frame's rendering time. For instance, functions that are more resource-intensive take a bigger percentage.

  - **Physics Frame %** is the same as **Frame %** but relative to the physics frame process.

- The **Time Scope** drop-down menu allows us to change the functions' time scope that we want to measure (highlighted and marked with **5** in *Figure 10.5*), and it has the following options:

  - **Inclusive**, which will take into account the time a function and all its nested functions took to render

  - **Self**, which will only take into account the individual time of each function without considering the function calls that the measured function made

- The **Frame #** stepper, or spinbox, marks the frame you are currently assessing (highlighted and marked with **6** in *Figure 10.5*). Changing the frame number will allow you to accurately see the function's measurements related to this frame in the **Functions** panel.

- The **Measurement Graph** panel is where the data is plotted so we can see it and access any unusual data. Each measured function has its own color to make it easy to see it on the graph (highlighted and marked with **7** in *Figure 10.5*).

The Profiler is a powerful ally ready to give us access to important data regarding resource management. Now that we understand how to use it, let's move on to the second profiler on the **Debugger** dock, **Visual Profiler**.

This one specializes in visual resources and potential bottlenecks so that we can improve our game's visuals regarding rendering and other visual procedures.

Exploring the Visual Profiler tab

On top of knowing how many processing resources your functions are taking from the CPU, it's also important to assess how much the rendering-related tasks (such as culling, lightning, and draw calls) are taking from the GPU. The **Visual Profiler** tool can help you keep track of what is causing the most delay in rendering a frame on the CPU and GPU. By identifying these sources of potential bottlenecks caused by rendering, you can optimize your CPU and GPU performance.

The **Visual Profiler** tab is quite similar to the **Profiler** tab but specializes in tracking and measuring rendering-related tasks. Take a look at the following figure to understand how the **Visual Profiler** tab displays all of this information.

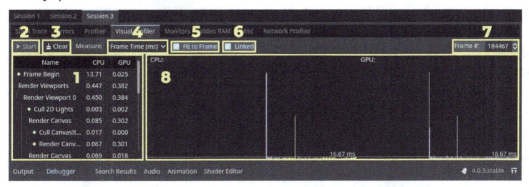

Figure 10.6 – Session 3's Debugger dock Visual Profiler tab and its elements

In order to gain a deeper understanding of each of these elements, let us take a closer look at their individual roles. Again, we are going to follow the elements in the numeric order:

- The **Tasks** panel displays the rendering-related tasks divided into categories. Note that they are broken down into elements such as the related viewport and canvas layer (highlighted and marked with **1** in *Figure 10.6*).

- The **Start** button, just like in the **Profiler** tab, initializes the profiling. Visual profiling is also turned off by default (highlighted and marked with **2** in *Figure 10.6*).

- The **Clear** button clears the current data gathered in the profiling session (highlighted and marked with **3** in *Figure 10.6*).

- The **Measure** drop-down menu (highlighted and marked with **4** in *Figure 10.6*) allows us to select two measurement options:

  - **Frame Time (ms)** is the time taken to render a frame in milliseconds

  - **Frame %** is the percentage a given procedure takes from the rendering time of a given frame

- The **Fit to Frame** checkbox will fit the graph to the default frame scale (highlighted and marked with **5** in *Figure 10.6*). Disable it to fit the graph onto over 60 **Frames Per Second** (**FPS**) portions.

- The **Linked** checkbox zooms the CPU and the GPU graphs to fit the same scale (highlighted and marked with **6** in *Figure 10.6*).

- The **Frame #** stepper, just like in the **Profiler** tab, marks the current frame you are assessing. Rendering tasks displayed in the **Tasks** panel relate to this frame (highlighted and marked with **7** in *Figure 10.6*).

Visual Profiler is yet another powerful ally when optimizing the rendering performance of your game and is a game-changing tool that can help you assess what may be causing lags and frame drops in your game. In the next section, we are going to understand yet another powerful tool available for us to assess our game's health, the **Monitors** tab, where we can find all sorts of interesting information regarding our game.

Well, let's dive into it so we understand how each of these, among other available data, will help us address potential issues in our game's performance.

Exploring the Monitors tab

Here is the part where you can really feel like a game doctor. **Monitors** allows us to assess important data as graphs and see the game's overall health. In this tab, we can track performance-related data in graphs. By default, it presents some useful data, such as the following:

- Time-related data, such as **FPS**, process time, and physics process time

- Memory-related data, such as static memory, dynamic memory, and message buffers

- Object-related data, such as the total object count, resource count, node count, and orphan nodes

There's a series of properties you can track and plot into graphs so you can analyze your game's health and spot potential areas for improvement. In the following figure, you can see the Debugger **Monitors** tab with some properties being tracked and plotted. Note that these properties are toggled on in the left panel by default. The **Monitors** tab will only plot charts for properties we toggle on in the left panel:

Figure 10.7 – Session 2's Debugger dock Monitors tab and its elements

The **Monitors** tab seems to be the simplest of the tabs we've seen so far, but it is still very powerful, so let's understand the two core elements that build it up:

- The **Monitor** panel is where we can find the available monitors. A monitor is data marked for tracking. Note that there are plenty of monitors by default. By using them, we can obtain some valuable information about our project's health (highlighted and marked with **1** in *Figure 10.7*).

- The **Graphs** panel is where the monitors are plotted as graphs, and each monitor has its own graph and measures. Only the monitors checked in the **Monitor** panel are plotted in the **Graphs** panel (highlighted and marked with **2** in *Figure 10.7*).

Note that there is no *start*, *stop*, or *clear* button on the **Monitors** tab. This is because Godot will always track the monitorable data.

Something interesting about the **Monitors** tab is that, on top of the default monitors, we can also create custom monitors using the `Performance` singleton. We are going to talk about that in the *Identifying the project's bottlenecks* section, where we will also talk about the **Monitors** panel in depth. In the next section, we will talk about the **Video RAM** tab, where we can assess our video-related resources.

Getting to know the Video RAM tab

The **Video RAM** tab is useful when you want to understand what resources are causing the most impact on your video memory. This is of great help, especially in 3D games, but it can also be useful for 2D games – for instance, when we want to assess whether we need to pack more sprites into a single texture.

The **Video RAM** tab is quite a simple panel with the essential information you need to assess video-related memory consumption. In the following figure, we can see it is made of a single table with four columns inside a panel:

Figure 10.8 – Session 2's Debugger dock Video RAM tab panel

This is an intuitive panel with the necessary data we need to understand the video memory usage of our resources. Let's understand the type of information each of these columns presents:

- **Resource Path** is the path in our Godot Engine's project to the resource.

- **Type** is the type of the tracked resource. In the preceding figure, all the resources are of the **Texture** type, but this can be a variety of other types. It can be useful to compare whether it's worth making an `AtlasTexture` Resource or a set of simple textures, for instance.

- The **Format** column is where we can find the data regarding the file's format.

- **Usage** is what we actually want in the end. It answers an important question: Given all the previous information, how much memory does this resource take?

> **Note**
>
> There's an option to save the table as a CSV file if you want to export it and make some table operations or create charts. This can be very useful for presentations, for instance.

The **Video RAM** tab in Godot Engine's **Debugger** dock is useful for assessing video-related memory consumption in games. It presents information on the resource path, type, format, and usage of video memory for each resource, helping us understand which resources we can improve, combine, remove, mix, and match to squeeze our available memory and open space for more resources. This kind of optimization is also good for low-end devices as, by using this, we can decrease the necessary specs to run the game. In the next section, we will talk about the **Misc** tab, where we can find the last clicked `Control` node during the debugging section.

Grasping the Misc tab

As mentioned previously, the Debugger **Misc** tab allows us to see the last `Control` node clicked in the **Debugging** section; note that this updates in runtime. This can be helpful to address crashes, screen flow, and most likely, which `Control` node is consuming inputs, and fix that in case we can have another `Control` node responsible for that. For instance, this is common when you have a `ColorRect` node that you use to fade the screen. If you don't set **Mouse Filter** to **Ignore**, it will consume mouse events and prevent the player from interacting with other UI elements. In the following figure, we have the **Misc** tab for our game:

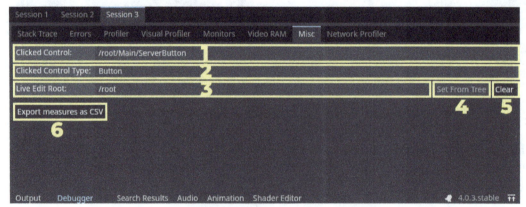

Figure 10.9 – Session 3's Debugger dock Misc tab and its elements

The **Misc** tab is quite simple, and we can't do a lot with it. Still, it's a good companion when we want to address interface-related issues, so let's understand the elements that build up this debugging tool:

- The **Clicked Control** row displays the last clicked `Control` node in the debugging section (highlighted and marked with **1** in *Figure 10.9*).

- The **Clicked Control Type** row displays the type of the clicked control (highlighted and marked with **2** in *Figure 10.9*).

- The **Live Edit Root** row displays the current root node in the live `SceneTree` instance (highlighted and marked with **3** in *Figure 10.9*).

- The **Set From Tree** button has no official documentation and it seems to be disabled all the time, so we couldn't test what this button does (highlighted and marked with **4** in *Figure 10.9*).

- The **Clear** button clears the data in the previously mentioned rows (highlighted and marked with **5** in *Figure 10.9*).

- The **Export measures as CSV** button allows you to export a CSV file with the data in the aforementioned rows. It might be useful for keeping track of how the game flows based on the interactions with its controls (highlighted and marked with **6** in *Figure 10.9*).

A good use case for this tab might be *Point 'n' Click* games. Since most interactions in this game happen with mouse clicks, we can use the Debugger **Misc** tab to identify which element led to a specific event. For instance, when clicking on a menu while a dialogue is being displayed, which one should consume the mouse click? Well, if the one you chose isn't consuming the input, you can use the Debugger **Misc** tab to see what's happening.

We just finished covering almost all the tools we can use to debug and profile our game. The only missing one is, for you, our fake studio's network engineer, the most important one. The Network Profiler is where you are going to find the impact of your RPCs and synchronizers, along with other relevant information related to the High-Level Network API. Let's get right into it!

Understanding the Network Profiler

It's time to meet your best ally, the one that will help you address issues related to your craft as the network engineer of our fake studio and come up with potential solutions for the problems that appear along your journey. The Network Profiler, as the name suggests, is a profiler specialized in network-related profiling. It displays information about RPCs' size and count, both sent and received, the node making and receiving the RPCs, `MultiplayerSynchronizer` nodes' network consumption and syncing count, and even a bandwidth meter, which are all we need to assess the impact of our network code.

Note that the Network Profiler, by default, only tracks the High-Level Network API bandwidth. So, if you are using low-level approaches, such as `PacketPeerUDP`, `UDPServer`, `StreamPeerTCP`, and `TCPServer`, their consumption may not be taken into account by the Network Profiler by default. We are going to see how we can address that in the *Using the Monitors tab* section.

Let's dive into the features we have available in the **Network Profiler** tab. Again, each element in this interface will be numbered for further reference.

Figure 10.10 – Session 2's Debugger dock Network Profiler tab and its elements

Although the Network Profiler has fewer elements than the other profilers, each of its elements is more complex as well. You may have also noticed that there's no graph element, right? So, assessing this data can be a bit less natural. But let's understand what each of these elements does and how we can use them:

- The RPC panel displays each node that sends and receives RPCs (highlighted and marked with **1** in *Figure 10.10*). On the right, we can see the **Incoming RPC** count and size and the **Outgoing RPC** count and size in bytes:

  - The **Node** column displays the node path to the node sending and receiving RPCs.

  - The **Incoming RPC** column displays the count and size of RPCs this node is receiving in bytes. This means other nodes in the network are calling RPCs in this particular node. Note that all `Asteroid` nodes have an **Incoming RPC** value of 3, which is probably because they receive 3 calls to process damage and are destroyed right after the third one.

  - The **Outgoing RPC** column displays the count and size, in bytes, of RPCs this node is sending. This means it is calling RPCs on other nodes. Notice that `Weapon2D` has a big **Outgoing RPC** value because it's constantly telling its peers' instances to fire bullets.

- The **Start** button. Just like in the previous profilers, network profiling is turned off by default; by pressing this button, we can start profiling (highlighted and marked with **2** in *Figure 10.10*).

- The **Clear** button clears the current data gathered in the profiling session (highlighted and marked with **3** in *Figure 10.10*).

- The **Bandwidth** meter displays the total bandwidth consumption in bytes per second of the current profiling session (highlighted and marked with **4** in *Figure 10.10*):

  - **Down** displays how many bytes per second it downloaded in this profiling session

  - **Up** displays how many bytes per second it uploaded during this profiling session

- The **Synchronization** panel displays all the `MultiplayerSynchronizer` nodes, their `SceneReplicationConfig` resource, which is always built-in by default, the sync count, and sync size in bytes (highlighted and marked with **5** in *Figure 10.10*):

  - The **Root** column displays the root node of the `MultiplayerSynchronizer` node's scene

  - The **Synchronizer** column displays the tracked `MultiplayerSynchronizer`

  - The **Config** column displays the `SceneReplicationConfig` resource associated with `MultiplayerSynchronizer`

  - The **Count** column displays how many times this particular `MultiplayerSynchronizer` node synced its replication data

  - The **Size** column displays the total amount of data the synchronization took, in bytes, during the current profiling session

With all this information in our hands, we can make sense of how our work is having an impact on the project's overall performance. Knowing how many times a node calls its RPCs, how many times other nodes call its RPCs, the amount of data exchanged, and more can help us properly address the necessary bandwidth a player needs to play the game properly, and also optimize the game to embrace players with lower network profiles.

In the next section, we are going to learn how we can use the powerful tools we've seen so far to spot the bottlenecks in our network approach using the Network Profiler, but also extend our profiling by adding custom monitors to the **Monitors** tab. With that, we can pinpoint what we want Godot to report to us.

Identifying the project's bottlenecks

With all the tools we've seen so far in this chapter at our disposal, it's time to use them to assess our project's health and look for areas of improvement. Since your focus here is on networking, we are going to concentrate on features related to this area. In this section, we will use the final version of the *Chapter 9, Creating an Online Adventure Prototype*, project to look for areas of improvement using **Network Profiler** and the **Monitors** debugging tools. You will learn how to do the following:

- Analyze the incoming and outgoing RPC count and size to identify potential bottlenecks in the network code

- Use the bandwidth meter to track the total bandwidth consumption and come up with possible solutions

- Assess the synchronization count and size of `MultiplayerSynchronizer` nodes to optimize replication data

- Create custom monitors to analyze relevant data specific to your project and track potential issues

Let's get started with the tool that will be our reliable companion through the process of identifying the issues related to the High-Level Network API, the Network Profiler.

Using the Network Profiler

In the previous section, we saw the Network Profiler, one of the most powerful tools available to us for identifying issues related to the High-Level Network API. In this section, we will dive deeper into using the Network Profiler to identify bottlenecks related to RPCs and `MultiplayerSynchronizer` nodes. To accomplish this, we will be using the final version of the *Chapter 9, Creating an Online Adventure Prototype*, project.

As mentioned before, we can use the Network Profiler to gather information about the size and count of a node's incoming and outgoing RPCs, `MultiplayerSynchronizer` nodes' network consumption, syncing count, and even a bandwidth meter. By understanding and analyzing this data, we can identify potential issues in our network code and come up with possible solutions.

To start, let's take a closer look at the incoming and outgoing RPC count and size to identify potential bottlenecks in our network code. We will also use the bandwidth meter to track the total bandwidth consumption and come up with possible improvements.

After that, we will assess the synchronization count and size of `MultiplayerSynchronizer` syncing to optimize replication data.

By the end of this section, you will have a better understanding of how to use the Network Profiler to identify and address issues related to your game's network performance. So, let's get started!

RPCs are a straightforward and efficient way to pass data and trigger remote events over the network. However, it's important to use them judiciously to avoid overloading the network.

In this section, we will analyze the data related to our project's RPCs and explore potential improvements. We will be implementing solutions in the next chapters, but for now, our focus is on learning how to look at the data critically and make informed decisions.

We start by playing the `res://09.prototyping-space- adventure/MainMenu.tscn` scene with three game sessions opened. Let's start the Network Profiler on all three of them.

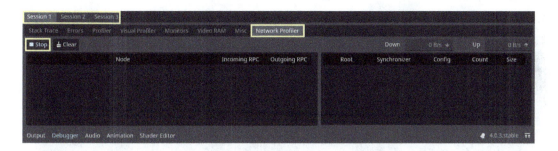

Figure 10.11 – The Session 1 Debugger Network Profiler starting profiling

Then, let's pick one to be the server while using the others as clients, in other words, as players. To enable multiple game sessions, you can select the **Run 3 Instances** option in the **Debug → Run Multiple Instances** menu.

With all three sessions opened, let's identify which one is the server. For that, open the Debugger **Misc** tab and look for the one where the last **Clicked Control** is `ServerButton`. In my case, it's the **Session 2** game instance, as shown in the following figure.

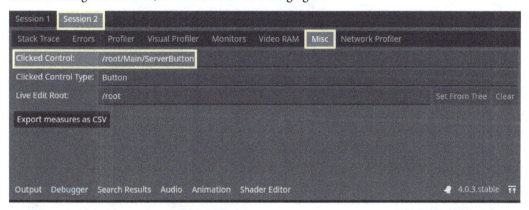

Figure 10.12 – Using Session 2's Debugger Misc tab to find the server's game instance

Now that we know that **Session 2** corresponds to the server, we can infer that the other two sessions are the clients. With this information in mind, we can start debugging with some premises. For instance, we can expect that the server will have a higher count of RPCs, especially the server's `QuestDatabase`.

To test whether this modification worked, I destroyed the *Asteroids* with one of the players' game instances, so go ahead and do the same. After destroying all 30 *Asteroids*, let's analyze the data that the Network Profiler collected. At this point, you can stop the Network Profiler if you want. In the following figure, we have the **Session 1** data, so we can assume it's a client.

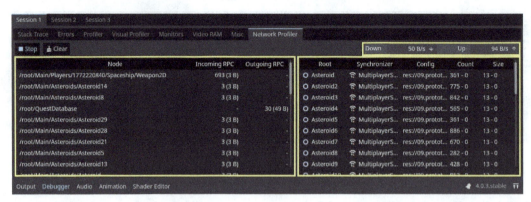

Figure 10.13 – The Session 1 Debugger Network Profiler displaying the data it gathered

Let's start with a brief analysis of the RPC counts and size. You can see that in the first row, we have a player `Spaceship/Weapon2D` with a high **Incoming RPC** count. This is probably the `fire()` method being called from another client's game instance, so we can assume the player that destroyed the *Asteroids* used **Session 3**.

This client called this method 693 times. The `fire()` method doesn't rely on any data that would require streaming it this many times. `Weapon2D` essentially has two major states:

- Firing
- Not firing

This means that we could improve this RPC count by sending a Boolean value through the network once when the player presses the firing action and when they release the firing action. In the meantime, `Weapon2D` itself would just toggle between these two states, firing and not firing, and use `process()` to spawn *Bullets* based on their fire rate. This would reduce this RPC count by a lot.

Did you notice how important this assessment can be as you build your project and adjust it along the way? Pretty cool, right?

Next, let's take a look at the `QuestDatabase` node in the fourth row. It's the only node that has an outgoing RPC count, right? So, it's making requests to the server's game instance. It made a total of 30 RPCs, but notice that their size is comparatively bigger than the 693 incoming `fire()` RPCs. This means that the data transmitted through this RPC is bigger. We should pay attention to it. This is likely to be the `update_player_progress()` method. Notice that we have 30 *Asteroids*, and every time we destroy one of them, we make an RPC to the `update_player_progress()` method. The count is correct and I can't see a clear area of improvement in this regard. It has a ratio of 1:1 – one event, one trigger. So, we are likely to figure out a way to improve the data; maybe compress it somehow to decrease the overall bandwidth.

Finally, let's take a look at the *Asteroids*' RPC count. Every single one of them receives only 3 RPCs; this is probably due to the server's *Bullets* hitting the *Asteroids*, which leads to the server calling the `hit()` method twice on the clients' instances. Then, it calls the `explode()` method the third time a *Bullet* hits

the *Asteroid*. It seems that this class is pretty healthy regarding its RPC counts on the client's side. There's nothing to improve on this side of the relationship. Let's take a look at the server's side. The following figure showcases the server's Network Profiler. Note that, in this playtest, the server is represented by the **Session 3** instance of the game.

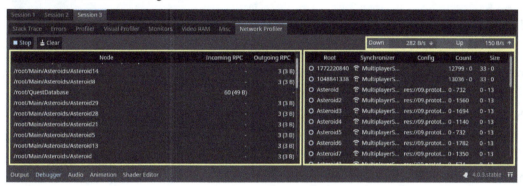

Figure 10.14 – The Session 3 Debugger Network Profiler displaying the data it gathered

The server has outgoing RPCs on its *Asteroids* instances, and they also have a total count of **3**, so it sounds like there's a symmetry here, right? You would be fooled into thinking that there are no improvements to make on the server side as well. Remember, in *Chapter 9's Separating server and client responsibilities* section, the server should calculate the damage and remove the *Asteroids* from the `SceneTree` instance, so there's no reason to play animations on the server. Ideally, the server will be a headless instance, so there's really no reason to play animations on it. But we won't dwell in these lands yet. What we can do in the current project is to change the `hit()` method's RPC annotation to only call it remotely and not locally. This way, at least the hit animation will only play on the client's side.

Let's take the chance that we are working with the *Asteroids* and also make an analysis of their `MultiplayerSynchronizer` nodes. You can see on the right panel's **Count** column that we have some quite high numbers. But remember… the *Asteroids* don't move, yet. At least not in this implementation. So why keep updating their positions constantly? The only time their properties should be synced is when the `World` node calls its `sync_world()` method. After that, there's no reason to keep updating the asteroids' properties. So, we can use the asteroids `MultiplayerSynchronizer` `update_visibility()` method inside the `sync_world()` method and decrease this bandwidth consumption as well.

By using the Network Profiler, we already identified areas for improvement, such as reducing the number of RPCs sent to the `Weapon2D.fire()` method and manually calling the `MultiplayerSynchronizer` syncing to decrease overall bandwidth. We also saw that we can change the RPC annotation of `Asteroid.hit()` to only call it remotely and not locally to reduce unnecessary animations on the server side.

Well, with just a brief analysis, we spotted some clear areas for improvement, didn't we? And we haven't even finished our assessment yet! In the next section, we will see how we can use the `Performance` singleton to create custom monitors and track them in the monitors track.

Using the Monitors tab

In the previous section, we learned about the Network Profiler and how it can help us identify potential bottlenecks in our game's network performance. In this section, we will focus on another powerful debugging tool in Godot Engine: the **Monitors** tab.

The **Monitors** tab allows us to track and analyze specific data points in real time. We can use it to keep track of variables, functions, and even custom data points that we define ourselves. By monitoring these data points, we can gain insight into how our project is performing and identify areas for improvement.

In addition to the built-in monitors, we can also create custom monitors to track specific variables or functions in our project. To do this, we need to use the `Performance.add_custom_monitor()` method, passing an ID, a `callable` instance, and, optionally, an array as arguments. Godot will create a monitor in the **Monitors** tab using the `id` argument and track the data using the `Callable` instance passed in the `callable` argument. This means that every time we trigger an event that should count for the data tracking, we need to execute the `callable` instance.

During the current section, we will use the **Monitors** tab to track some data regarding the `QuestDatabase` node and the `QuestSingleton` node. By monitoring these data points, we will gain some insights into how our quest system is performing and identify potential areas for improvement.

Let's start by opening the `res://09.prototyping-spaceadventure/Quests/QuestDatabase.gd` script. We are going to create a member variable to keep track of how many times the `QuestDatabase.update_player_progress()` method was called. We can name this variable `quest_update_count` and set its default value to 0. Then, we need to create a method that returns its current value; let's call this method `get_quest_update_count()`:

```
func get_quest_update_count():
    return quest_update_count
```

To update `quest_update_count`, let's increment its value after the server successfully updates the player's progress in a given quest. So, in the `update_player_progress()` method, add a line inside the `if multiplayer.is_server()` statement incrementing `quest_update_count` by 1:

```
@rpc("any_peer", "call_remote")
func update_player_progress(quest_id, current_amount, completed,
user):
    if multiplayer.is_server():
        progress_database[user][quest_id]["progress"] = current_
amount
        progress_database[user][quest_id]["completed"] = completed
        quest_update_count += 1
```

With that, we have everything ready to add `get_quest_update_count()` to our **Monitors** tab. For that, in the `_ready()` callback, create a `Callable` variable pointing to `QuestDatabase` using the `self` keyword, and pointing to `"get_quest_update_count"`. We can name this `Callable` variable `callable` to simplify the process:

```
func _ready():
    if multiplayer.is_server():
        load_database()

        var callable = Callable(self, "get_quest_update_count")
```

Then, let's call the `Performance.add_custom_monitor()` method. To keep things organized, we will use a category named `"Network"` for our custom monitors. So, in the `id` argument, we will pass `"Network/Quests Updates"` and pass `callable` as the second argument:

```
func _ready():
    if multiplayer.is_server():
        load_database()

        var callable = Callable(self, "get_quest_update_count")
        Performance.add_custom_monitor("Network/Quests Updates",
callable)
```

Let's start by opening the `res://09.prototyping-space- adventure/Quests/QuestDatabase.gd` script. We are going to create a member variable to keep track of how many times the method was called.

Now, to test whether this custom monitor is working and assess the data it will provide, let's test the game using three debugging sessions and use one of the clients to destroy some *Asteroids*. This time around, my server is on **Session 2**. The following figure showcases **Session 2**'s **Monitors** tab. You can find the **Quests Updates** monitor at the very bottom of the **Monitor** panel; tick the checkbox and Godot will display the tracked data.

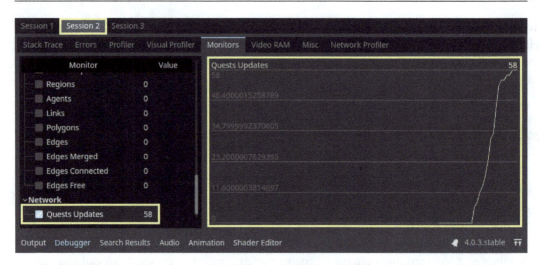

Figure 10.15 – The Session 2 Debugger Monitors tab displaying the Quests Updates tracked data

Notice that Godot counted only 58 quest updates. So I may have missed one asteroid. Interesting, right? Why does one *Asteroid* trigger two calls to the `QuestDatabase.update_player_progress()` method? Well, remember that currently the quest progress is shared among all peers, so this can grow exponentially. If there were 3 players, there would be 96 calls to `QuestDatabase.update_player_progress()`. We need to figure out a way to limit that. One quick solution is to check whether the quest is already completed, and if it is, stop updating it. This would limit this particular quest to 10 calls per player, which would be a good improvement.

Let's make this comparison, just for testing purposes. Open `res://09.prototyping-space-adventure/Quests/QuestSingleton.gd` and let's create a monitor that would only increment until the quest reaches the target amount required to complete the quest. To do that, let's create a new member variable called `increase_count` and set its value to 0 by default:

```
var increase_count = 0
```

Then, let's create a method called `get_quest_increases()` that will return this variable:

```
func get_quest_increases():
    return increase_count
```

In the `_ready()` callback, if this is a client instance, we will add a new custom monitor using the previous method as `callable`, just like we did with `QuestDatabase.get_quest_update_count()`:

```
func _ready():
    if not multiplayer.is_server():
        var callable = Callable(self, "get_quest_increases")
```

```
    Performance.add_custom_monitor("Network/Quest Increases",
callable)
```

Now, inside the `increase_quest_progress()` method, we will create an `if` statement that will only increment `increase_count` while `quest.current_amount` is less than `quest.target_amount`:

```
func increase_quest_progress(quest_id, amount):
  if not quest_id in quests.keys():
    return
  var quest = quests[quest_id]
  quest.current_amount += amount
  QuestDatabase.rpc_id(1, "update_player_progress", quest_id, quest.
current_am
out, quest.completed, AuthenticationCredentials.user)
  if quest.current_amount < quest.target_amount:
increase_count += 1
```

Let's test the game again and see what happens in the clients' **Monitors** tab. In the following figure, there's something very interesting happening.

Figure 10.16 – The Session 2 Debugger Monitors tab displaying the Quests Increases tracked data

This time around, **Session 2** corresponds to a player – particularly, the player logged in using the `user2` credentials. Why this is relevant? Notice that there were only four increments to `increase_count` in this game instance. This is due to the fact that, in the `res://09.prototypingspace-adventure/Quests/QuestDatabase.json` file, `user2` already destroyed five asteroids, so it only needed five more to complete the quest. This means that we can improve this aspect of our game even between play sessions. The more progress a player makes in a game session, the fewer RPCs we will need to make to the server if we implement this approach; pretty cool, isn't it?

Throughout this section, we learned how we can use the `Performance` singleton to create new monitors in the **Monitors** tab using the `Performance.add_custom_monitor()` method. We also saw how we can create methods to collect data about potential bottlenecks in our game. Finally, we saw some potential fixes to the issues we found while debugging the game in order to optimize it.

Summary

With that, we conclude our chapter! Throughout this chapter, we introduced the **Debugger** dock, which is a powerful tool for assessing and debugging potential problems in our game, as well as for optimizing its performance.

We explored the **Stack Trace** tab, which gives us an overview of our game's flow and provides us with multiple ways to gather information about the changes that happen throughout this flow, allowing us to understand the whole chain of cause and effect that led to a given change. We also talked about the **Errors** tab, which is where we work together with thousands of other developers who worked on the development of Godot Engine's core and identified thousands of errors and documented them so that when they happen, we have some light on the issue and can fix it.

On top of that, we explored two powerful performance-based debugging tabs: the **Profiler** tab and the **Visual Profiler** tab. The **Profiler** tab is one of your best allies in this task, as it allows you to see rendering time, physics simulation time, audio processing time, and even how much time each of your custom script functions is taking to process and how many times they were called. The **Visual Profiler** tab specializes in tracking and measuring rendering-related tasks and can help you keep track of what is causing the most delay in rendering a frame on the CPU and GPU.

However, the protagonists of the chapter were the **Network Profiler** and **Monitors** tabs. We saw how we can identify potential bottlenecks related to RPCs and `MultiplayerSynchronizer` nodes by analyzing data gathered by the tool. By understanding and analyzing this data, we came up with possible solutions to optimize network code. In addition to that, we learned how to use the `Performance` singleton and create custom monitors to track specific data points in real time in the **Monitors** tab. By monitoring these data points, we gained insights into how our project is performing and even made a test for a potential improvement.

In the next chapter, we are going to optimize data requests, especially regarding the quests data from the `QuestDatabase.get_player_quests()` method.

By optimizing the way we request and handle data, we can improve our game's performance and provide a better experience for our users. See you there!

11

Optimizing Data Requests

Welcome to *Chapter 11*, *Optimizing Data Requests*, where we will use the tools that we saw in *Chapter 10*, *Debugging and Profiling the Network*, and finally implement improvements to the network code that we wrote in *Chapter 9*, *Creating an Online Adventure Prototype*.

In this chapter, we will understand a bit more about bandwidth and throughput by analyzing the current state of our game. We saw, in the *Using the network profiler* section of *Chapter 10*, *Debugging and Profiling the Network*, that we have some improvements to make, especially regarding the MultiplayerSynchronizers and the QuestDatabase data transmission. So, here, we will see how we can decrease the number of requests and how we can compress and decompress data to reduce the bandwidth and throughput and make our game available to more people in a more reliable and optimal way.

By the end of this chapter, you will understand that there are many ways to optimize a game and most of the optimizations will depend on the specific demands of the game itself. As you progress, you will develop a keen sense and understanding of how to assess potential areas of improvement and the kind of data you are looking for, as well as some general techniques to address bottlenecks in your network code.

So, this chapter will cover the following topics:

- Understanding network resources
- Decreasing the requests count
- Data compression with ENetConnection

Technical requirements

As mentioned in *Chapter 10*, *Debugging and Profiling the Network*, *Part 3*, *Optimizing the Online Experience*, is based on the final version of the project made in *Chapter 9*, *Creating an Online Adventure Prototype*, so it's fundamental to have read, practiced, and implemented the concepts presented there.

You can get the files necessary to get started with this chapter through the following link. These files have the implementations we've made in *Chapter 10, Debugging and Profiling the Network*:

```
https://github.com/PacktPublishing/The-Essential-Guide-to-Creating-
Multiplayer-Games-with-Godot-4.0/tree/11.optimizing-data-requests
```

It's also necessary that you have read and understood the concepts and tools presented in *Chapter 10, Debugging and Profiling the Network*, so that we can move forward under the assumption that you already know what they are and how to use them properly. In particular, you will need to understand how the Debugger's Network Profiler and Monitors tools work. So, if you aren't sure how to use these tools, please take some time to go back and read *Chapter 10, Debugging and Profiling the Network*, so you can master these and other debugging tools.

Understanding network resources

We've already mentioned the importance of bandwidth and throughput; in *Chapter 1, Setting up a Server*, we even had a brief introduction to and a visual representation of the topic in *Figure 1.3* and *Figure 1.4*. Now it's time to wrap our heads around these concepts, which are fundamental to network usage optimization and will be our major resources to measure the improvements we made toward our optimization goals.

As a general rule, the less bandwidth and the lower the throughput of our network code, the better. Of course, we need to keep in mind that all optimizations should maintain the game experience, so we are in a very delicate position. Different from other processing, memory, and graphics optimizations, our work can't create "beautiful accidents," such as a processing optimization that can lead to a cool mechanic. No, our job as network engineers is to replicate the already established mechanics and effects to all peers in a network. Let's understand how we do that.

When talking about bandwidth, it might be surprising to hear that most games don't actually need to have huge infrastructures available. For instance, a video conference takes way more bandwidth than a complex first-person shooter or a war simulator with tons of physics simulation because it works with processed data in the form of rendered images that must be passed around and re-created in each of the participants' instances of the conference. In the case of games, most of the necessary resources to create the simulations are already available on the user's machine, so most of our job is to communicate through messages what the computer should load to sync the client's and server's game instances. It's well established that players need a bandwidth of about 5 Mbps to play most modern online multiplayer games, including huge franchises such as Call of Duty, League of Legends, and Fortnite.

In *Part 1, Handshaking and Networking*, of this book, we saw that games work mostly with unreliable data, meaning that most of the time, we only need to know the latest data regarding an object in the game in order to sync it. This decreases the network usage a lot and allows us to focus on the specific types of data necessary to replicate the server's game world in clients' game worlds.

The major concern in online multiplayer games is the consistency with which we can keep on the data transmission stream – ultimately, whether we can maintain the throughput of our network consistently. This can be affected by latency and other external aspects, so all we can do is design a communication architecture that takes into account how latency can affect our throughput. Of course, we will also try to keep the bandwidth to a minimum so that if a household has many devices connected to the same network, our game has room to keep the data flowing.

So, keep in mind that bandwidth and throughput are our major resources and we will be looking at them to find areas that we can improve in our game.

You might still be trying to work out the difference between bandwidth and throughput. So, let's briefly assess them both. We were introduced to these concepts in *Chapter 1, Setting up a Server*, so if you don't quite remember what they are, take a brief moment to read the *What is the UDP protocol?* section of the chapter.

We are going to use bandwidth to understand how much of the network our game needs to perform correctly, meaning how much data we expect to transmit through the network taking into account all the measurements we observed using the network profiler. This means that if we have 1,000 MultiplayerSynchronizers syncing 5 KB of data per second at any given point of our game, we will need a network with a 5 Mbps velocity. Note that this might not be consistent transmission of 5 Mbps throughout the whole game session, but it's a recommendation based on our measurements that the game might require a network that can manage up to 5 Mbps to play smoothly. In summary, bandwidth is the amount of space available in a connection for data transmission, not the transmission itself.

Now, the actual thing we are looking to optimize is the **throughput**, which points to how many packets and the size of these packets we are actually transmitting through the network. The throughput is our actual data flow. You can think about it using an analogy where bandwidth is a pipe and throughput is the water. We can't stream more water than a pipe can support; instead, we can stream water up to the available pipe size. In the same way, the amount of throughput we can have is based on the available bandwidth capacity.

You can see an illustration of a good and bad throughput-bandwidth relationship in the following figure:

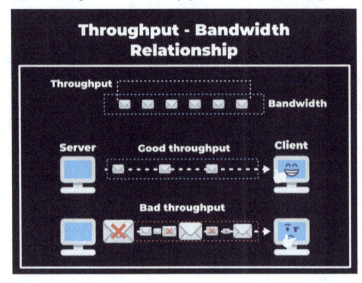

Figure 11.1 – An illustration of a good and bad throughput-bandwidth relationship

Note that with good throughput, the data sent and received is consistent and doesn't exceed the bandwidth, whereas with bad throughput, the data packets are inconsistent in size and frequency, and some are even lost in the middle of the transmission. When packets don't reach their destination, we call it **packet loss**, and this can cause lots of headaches and complaints from players.

With packet loss, the client doesn't know how to handle their game instance properly. Where should the second player's spaceship be? Did it stop shooting? Is it shooting at an `Asteroid` node or are there no more `Asteroid` nodes to shoot? Is the `Asteroid` node they were shooting at there yet? We are going to see, in *Chapter 12, Implementing Lag Compensation*, how to handle these situations, but ideally, we should avoid them by paying attention to the throughput.

With all that said, network engineers are left in a very delicate position when the resources available for optimization don't leave space for new mechanics to emerge. We take the already implemented mechanics and try to fit them all into the available resources, usually squeezing them to keep up with any potential changes that may require more of these resources, for instance, a new mechanic that requires more bandwidth. So, we can't experiment a lot when optimizing the network.

In the next section, let's talk about how we can initiate our optimizations by decreasing the number of requests that our game makes. We are going to assess the problems presented in *Chapter 10, Debugging and Profiling the Network*, such as in `Weapon2D` massive RPC count and unnecessary syncs of the `Asteroid` node's `MultiplayerSynchronizer` node.

Decreasing the requests count

Previously, in *Chapter 10, Debugging and Profiling the Network*, we saw that there was a disproportional and unnecessary number of requests being made to the Weapon2D node's fire() RPC method, and we even came up with what could be a solution for this issue. We also saw that we can decrease the Asteroid node's sync problem by only updating it once a given player requests a synchronization using the World node's sync_world() RPC method.

In this section, we are going to implement these optimization methods and improve the overall performance of our network. Let's start with the Weapon2D node issue.

Reducing the weapon fire count

Sometimes we may need to make changes in the core code of a feature in order to improve its network performance, even when its local performance stays the same, or may even drop. When talking about optimization, we are always trying to balance things out and figure out how to use the available resources in a way that allows more players to enjoy a better experience. Network resources are particularly a priority in most online multiplayer games since it's through a good network that players will be able to get the most out of their shared experiences. So, let's make some changes to how Player2D node fires Weapon2D node. Let's open the res://09.prototyping-space-adventure/Objects/Weapon/Weapon2D.tscn scene first and carry out the following steps:

1. Connect the timeout signal of Timer node to Weapon2D node and create a callback method named _on_timer_timeout():

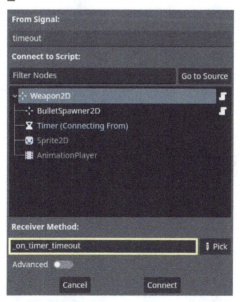

Figure 11.2 – Timer node's timeout signal connecting to

theWeapon2D node's _on_timer_timeout() callback

2. Open the `res://09.prototyping-space-adventure/Objects/Weapon/Weapon2D.gd` script and, in the `_on_timer_timeout()` callback, call the `fire()` method so `Weapon2D` node fires on every `Timer` node's tic.

3. Then, let's create an RPC method that can be called by any peer and should also be called locally. We will use this method to change `Weapon2D`'s firing state, so it should receive a Boolean variable as an argument:

```
@rpc("any_peer", "call_local")
func set_firing(firing):
```

4. Inside this method, we will check the `firing` state, and if it's `true`, we will call the `fire()` method as well; otherwise, we will tell `Timer` to stop:

```
@rpc("any_peer", "call_local")
func set_firing(firing):
    if firing:
        fire()
    else:
        timer.stop()
```

5. With that, we can remove the RPC annotation on the `fire()` method and the `if timer.is_stopped()` statement, since now `Timer` itself tells when `Weapon2D` fires. The `fire()` method should look like this after that:

```
func fire():
    animation_player.play("fire")
    spawner.spawn(bullet_scene)
    timer.start(1.0 / fire_rate)
```

With that, `Weapon2D` will fire based on the timeout signal from `Timer`. With the new RPC method, we can change the firing state, starting or stopping the creation of new *bullets*. The `Weapon2D` script should look like this at this point:

```
class_name Weapon2D
extends Marker2D

@export var bullet_scene: PackedScene
@export_range(0, 1, 1, "or_greater") var fire_rate = 3
@onready var spawner = $BulletSpawner2D
@onready var timer = $Timer
@onready var animation_player = $AnimationPlayer
func fire():
    animation_player.play("fire")
    spawner.spawn(bullet_scene)
    timer.start(1.0 / fire_rate)
```

```
@rpc("any_peer", "call_local")
func set_firing(firing):
  if firing:
    fire()
  else:
    timer.stop()
func _on_timer_timeout():
  fire()
```

Now, we need to know when Weapon2D's firing state changes, and to do that, we will need to make some changes to Player2D. So, open res://09.prototyping-space-adventure/Actors/Player/Player2D.gd and implement the following steps:

1. Remove the whole _process() callback code. Then, override the _unhandled_input() callback:

    ```
    func _unhandled_input(event):
    ```

2. Inside the _unhandled_input() callback, we are going to check whether the "shoot" action was pressed or released. If pressed, we set the weapon firing state to true, and if released, it's set to false (remember, we should do that using the rpc() method so the player shoots on all network peer instances):

    ```
    func _unhandled_input(event):
      if event.is_action_pressed("shoot"):
        weapon.rpc("set_firing", true)
      elif event.is_action_released("shoot"):
        weapon.rpc("set_firing", false)
    ```

3. Next, we need to add a line to the setup_multiplayer() method to also toggle the _unhandled_input() process based on whether the instance of the spaceship is the current player or a remote player:

    ```
    func setup_multiplayer(player_id):
      var self_id = multiplayer.get_unique_id()
      var is_player = self_id == player_id
      set_process(is_player)
      set_physics_process(is_player)
      set_process_unhandled_input(is_player)
    ```

With that, Player2D will toggle Weapon2D's firing state based on whether the "shoot" action was pressed or released, instead of calling it every frame while "shoot" was pressed.

Let's make an assessment of this improvement. Turn on the network profiler and let's see how this goes. Remember that depending on the duration of the profiling section, we may get different results, so this isn't as accurate as a unit test; but still, it will give us a good sense of any potential improvements we've made. In the following figure, we have **Session 1** as the server, receiving RPCs from the client's `Player2D` instance:

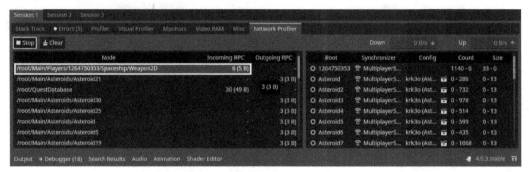

Figure 11.3 – The Session 1 Debugger's Network Profiler tab
highlighting thePlayer2D node instance's incoming RPCs

As you can see in the figure, I destroyed all 30 asteroids in a play session of around 20.0 seconds, with a total of six RPC calls from `Player2D` instance. If you compare this to *Figure 10.13*, where we had 693 calls for a session with about the same duration, this is more than 115x fewer calls. I think we did a good job here.

So, in this section, we learned how to optimize data requests in our game to improve network performance. We focused on how to decrease the number of requests being made in the game by optimizing the firing of `Weapon2D` node. We also saw how to create an RPC method that can change `Weapon2D`'s firing state, and how to toggle the firing state based on the `"shoot"` action being pressed or released instead of triggering the firing on every frame in the `_process()` callback. Finally, we saw how to use the network profiler to assess the impact of these optimizations on the game's performance. In the next section, we will work on decreasing the requests of Asteroid's `MultiplayerSynchronizer`.

Decreasing Asteroid's syncing count

Another issue we saw in *Chapter 10, Debugging and Profiling the Network*, was regarding the way `Asteroid` node's `MultiplayerSynchronizers` sync their position to players. Since they don't move throughout the play session, there's no need to keep updating their position and other properties on a regular basis. Instead, we only need to sync them once when the player asks `World` node to sync.

So, open the `res://09.prototyping-space-adventure/Objects/Asteroid/Asteroid.tscn` scene and make the necessary changes to improve this aspect of our network engineering.

Here, we just need to set the `MultiplayerSynchronizer` node's **Visibility Update Mode** property to **None**:

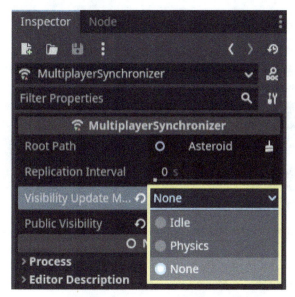

Figure 11.4 – Asteroid node's MultiplayerSynchronizer node Visibility Update Mode property set to None

With that, we just have to do the syncing manually now. The `MultiplayerSynchronizer` node has a method called `update_visibility()`, which receives an argument where we can pass the ID of the peer we want to sync with; if we pass 0, it updates all peers instead.

Note that this method takes into account the filters we set in the *Syncing the asteroids* section of *Chapter 9*, *Creating an Online Adventure Prototype*, using the `set_visibility_for()` method, meaning only the peers added using this method will be synced. So, in our case, if we don't add the peer to the filtering using the `set_visibility_for()` method, this peer won't be synced, even if we use the `update_visibility()` method, passing the correct peer ID.

The best place to manually sync the `Asteroid` node's properties will be in the `World` class, so open the `res://09.prototyping-space-adventure/Levels/World.gd` script. Then, inside the `sync_world()` method, let's add another group call to the `"Sync"` group, but this time to the `update_visibility` method, and pass `player_id` as an argument. The whole `sync_world()` method should look like this:

```
@rpc("any_peer", "call_local")
func sync_world():
    var player_id = multiplayer.get_remote_sender_id()
    get_tree().call_group("Sync", "set_visibility_for",
        player_id, true)
```

```
get_tree().call_group("Sync", "update_visibility",
    player_id)
```

With that, all `Asteroid` nodes should only sync once to every player, decreasing the total network resource usage.

Let's profile this change as well to see its impact on the overall network consumption. The following figure showcases the sync count of some asteroids' MultiplayerSynchronizers:

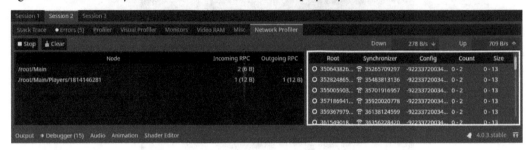

Figure 11.5 – The Session 2 Debugger's Network Profiler highlighting the
Asteroid nodes' MultiplayerSynchronizer nodes' sync counts

Compare this to *Figure 10.13*, where we had hundreds and hundreds of updates, and for every frame, the update count would increase. After what we've done, they only increase once a player enters a session and only to that specific player. Pretty good job, right?

In this section, we saw how to optimize data requests by decreasing the `asteroid` nodes' `MultiplayerSynchronizer` nodes' syncs. To do that, we disabled the automatic synchronization by setting `MultiplayerSynchronizer` node's **Visibility Update Mode** property to **None** and manually syncing by using the `MultiplayerSynchronizer.update_visibility()` method once the player joins the game.

In the upcoming section, we will see how we can compress data to optimize the packet's size. Until now, we've only dealt with how many packets we send or receive, but the size of these packets is very important to handle as well. Let's understand what we have available to do that.

Compressing data with the ENetConnection class

The Godot Engine's high-level network API includes the `ENetConnection` class. This class is available after a connection between peers, such as the server and client, is established. Using `ENetConnection`, we can tweak the peer's connections, allowing us to tell, for instance, the compression method. We can access the `ENetConnection` instance after the peers have successfully connected. For that, we can use the `ENetMultiplayerPeer.host` property.

There are five available compression methods to use in the `ENetConnection` class:

- **None**, which can be used through the `CompressionMode` enumerator's `COMPRESS_NONE` option. This is what the documentation says:

  This uses the most bandwidth but has the upside of requiring the fewest CPU resources. This option may also be used to make network debugging using tools like Wireshark easier.

- ENet's built-in **Range Coder**, which can be used through the `CompressionMode` enumerator's `COMPRESS_RANGE_CODER` option. This is what the documentation says:

  [This is] ENet's built-in range encoding. Works well on small packets, but is not the most efficient algorithm on packets larger than 4 KB.

- **FastLZ**, which can be used through the `CompressionMode` enumerator's `COMPRESS_FASTLZ` option. This is what the documentation says:

  This option uses less CPU resources compared to COMPRESS_ZLIB, at the expense of using more bandwidth.

- **ZLib**, which can be used through the `CompressionMode` enumerator's `COMPRESS_ZLIB` option. This is what the documentation says:

  This option uses less bandwidth compared to COMPRESS_FASTLZ , at the expense of using more CPU resources.

- **Zstandard**, which can be used through the `CompressionMode` enumerator's `COMPRESS_ZSTD` option. This is what the documentation says:

  Note that this algorithm is not very efficient on packets smaller than 4 KB. Therefore, it's recommended to use other compression algorithms in most cases.

You can find more about `ENetConnection` and `CompressionMode` in the Godot docs through the following link:

`https://docs.godotengine.org/en/stable/classes/class_enetconnection.html#enum-enetconnection-compressionmode`

It's time to work on the size of the data we are transmitting through the network. In the following steps, we will implement compression to decrease our game's packet size and optimize the transmission of data through the network. Note that, as always, there's a trade-off between the bandwidth and CPU resources. Our game currently doesn't have any issues regarding CPU usage. So, at this point, we can focus on optimizing the network resources. For that, we can use the COMPRESS_ZLIB compression mode. To do that, let's open the res://09.prototyping-space-adventure/Authentication.gd script and complete the following steps:

1. Right before we set multiplayer_peer to the peer in the _ready() callback, we are going to change the ENetConnection compression mode. For that, we access the host property and use the compress() method, passing EnetConnection.COMPRESS_ZLIB as an argument:

    ```
    func _ready():
      if multiplayer.is_server():
        peer.create_server(PORT)
        load_database()
      else:
        peer.create_client(ADDRESS, PORT)
        peer.host.compress(EnetConnection.COMPRESS_ZLIB)
      multiplayer.multiplayer_peer = peer
    ```

    We need to do this here because the compression mode needs to be set before the connection is established, which happens after we set the multiplayer.multiplayer_peer property.

2. We also need to do the same thing in the res://09.prototyping-spaceadventure/LoggingScreen.gd script so that this connection also matches the server's connection compression. Again, before setting multiplayer.multiplayer_peer, we set ENetConnection's compression to COMPRESS_ZLIB:

    ```
    func _ready():
      peer.create_client(ADDRESS, PORT)
      peer.host.compress(ENetConnection.COMPRESS_ZLIB)
      multiplayer.multiplayer_peer = peer
    ```

With that, we are able to change the compression mode used in our game's network connection as soon as the players join the game's world. For now, this won't make a huge impact. As we saw in the previous quotes from the documentation, most of the compression algorithms aim at having data packets be either below 4 KB or above 4 KB. Our game's packets currently haven't even reached the Kilobytes yet, so...this may not have a big, if any, impact at this point, to be honest.

If we want to measure how much bandwidth we are using and get at least a glance at any possible improvements, we can use the Debugger's **Monitors** to track EnetConnection instances' sent and received data. For that, we can use the ENetConnection.pop_statistic() method to create two relevant monitors using the Performance singleton to add our custom monitors. Let's do it:

1. Still in the World class script, create a method called get_received_data(). This method needs to return an integer or floating point so we can use it to create a monitor. In this case, it will return the received data statistic from the current ENetConnection. For that, we can use the pop_statistic() method, passing ENetConnection.HOST_TOTAL_RECEIVED_DATA as an argument:

```
func get_received_data():
    var enet_connection = multiplayer.multiplayer_
        peer.host
    var data_received = enet_connection.pop_statistic
        (ENetConnection.HOST_TOTAL_RECEIVED_DATA)
    return data_received
```

2. Then, we are going to create another method called get_sent_data() and do the same thing, but passing ENetConnection.HOST_TOTAL_SENT_DATA as an argument this time around:

```
func get_sent_data():
    var enet_connection = multiplayer.multiplayer_
        peer.host
    var data_sent = enet_connection.pop_statistic
        (ENetConnection.HOST_TOTAL_SENT_DATA)
    return data_sent
```

3. Now, in the _ready() callback, where we check to see whether this instance is the server or not, right above where we create the Asteroid instances, we are going to add the respective callable to the Performance singleton using the Performance.add_custom_monitor() method, like so:

```
var callable = Callable(self, "get_received_data")
Performance.add_custom_monitor("Network/Received
    Data", callable)
callable = Callable(self, "get_sent_data")
Performance.add_custom_monitor("Network/Sent
    Data", callable)
for i in 30:
    asteroid_spawner.spawn()
```

Now, we are able to monitor the differences between each compression mode and see the one that better fits our game. In the following figure, we compare the usage of the COMPRESS_ZLIB and COMPRESS_NONE compression modes:

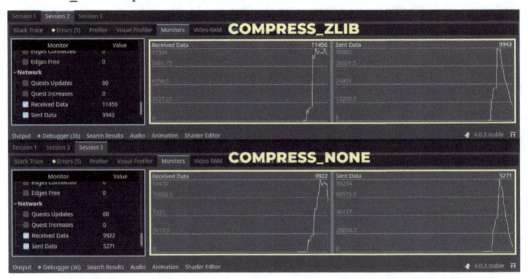

Figure 11.6 – A comparison between the monitors of the data sent and received
using the COMPRESS_ZLIB and COMPRESS_NONE compression modes

Note that with the COMPRESS_ZLIB compression, we had a peak of 12,509 bytes on the received data, while we had a peak of 48,802 bytes in the data sent. Meanwhile, using COMPRESS_NONE, we peaked at 14,470 bytes in the received data and 80,234 bytes on the sent data – even with very small data packets, we have massive gains, especially on the server's data sent metrics, so we also did a good job there!

Summary

In this chapter, we learned how to optimize data requests in our game to improve network performance. We focused on decreasing the number of requests being made in the game by optimizing the firing of the weapon. We saw how to create an RPC method that can change the weapon's firing state and how to toggle the firing state based on the "shoot" action being pressed or released instead of triggering the firing on every frame in the _process() callback. Finally, we saw how to use the network profiler to assess the impact of these optimizations on the game's performance.

After that, we tackled the issue of `Asteroid` node's MultiplayerSynchronizers' syncs. We disabled the automatic synchronization by setting the **Visibility Update Node** of `MultiplayerSynchronizer` to **None** and manually synced by using the `MultiplayerSynchronizer.update_visibility()` method once the player joins the game. This decreased the `Asteroid` node's sync count and decreased the total network resource usage. We also saw how to measure the effectiveness of these optimizations using the debugger's network profiler.

Finally, we learned about the Godot Engine's high-level network API's `ENetConnection` class, which offers compression methods to optimize the packet size. We saw how to use the `ENetConnection.pop_statistic()` method to create custom monitors using the `Performance` singleton to track ENetConnection's sent and received data. We compared the usage of the `COMPRESS_ZLIB` and `COMPRESS_NONE` compression modes and found that even with very small data packets, we have massive gains, especially on the server's data sent metrics.

In the next chapter, we are going to get even deeper into optimization. We are going to use interpolation and prediction to decrease the `MultiplayerSynchronizer` syncing count of the player's spaceship while trying to keep the movement consistent throughout the game network instances. See you there!

12

Implementing Lag Compensation

Welcome to one of the most anticipated chapters in the book. Here, we will dive into the core of online multiplayer game optimization. In the world of online gaming, where players from across the globe unite to embark on epic adventures, two formidable adversaries lurk in the shadows; they are **lag** and **latency**. These foes can transform a thrilling gaming experience into a frustrating trial. In this chapter, we'll confront these challenges head-on, arming you with the knowledge and tools to mitigate their impact and create an engaging online gaming environment.

In this chapter, we will use **Remote Procedure Calls** (**RPCs**) to implement lag compensation techniques, in order to make the `Player` node's `Spaceship` node maintain its position and rotation, synced throughout the game instances across the network. For that, we will understand the core issues regarding packet loss and latency, something common when we use unreliable packets, as we do when using the ENet protocols. Then, we will fake some latency and packet loss by using `Timer` nodes so that we can understand how these issues may display in the actual game. After that, we will talk about some common techniques to create solutions for these issues.

By the end of the chapter, you will understand how we can fake some smooth movement, even when the game's `MultiplayerSynchronizer` fails to deliver data across peers' game instances.

In this chapter, we will cover the following topics:

- Introducing lag issues
- Dealing with unreliable packets
- Common compensation techniques

Technical requirements

As mentioned in *Chapter 10*, *Debugging and Profiling the Network*, *Part 3*, *Optimizing the Online Experience*, of this book focuses on the final version of the project made in *Chapter 9*, *Creating an Online Adventure Prototype*, so it's fundamental to read, exercise, and implement the concepts presented there. You can get the files necessary to start this chapter at the following link: `https://github.com/PacktPublishing/The-Essential-Guide-to-Creating-Multiplayer-Games-with-Godot-4.0/tree/12.prediction-and-interpolation`. They contain the progress we've made for optimizations in *Chapter 11*, *Optimizing Data Requests*.

It's also necessary that you have read and understood the concepts and tools presented in *Chapter 11*, *Optimizing Data Requests*, so that we can continue with the assumption that you already know what they are and how to use them properly.

Introducing lag issues

Addressing lag and unreliable packets involves three techniques – **interpolation**, **prediction**, and **extrapolation**. These techniques smoothen player movements, maintain responsiveness, and anticipate object movements. The Godot Engine's physics simulation and RPC method are crucial in these techniques, aiding in realistic object movement and data synchronization, despite network issues.

Lag and latency are the archenemies of any online multiplayer game. Lag, often used interchangeably with latency, refers to the delay between a player's action and its corresponding effect in the game. It's the momentary pause between pulling the trigger and seeing the enemy fall. Latency, on the other hand, represents the time it takes for data to travel from a player's device to the game server and back. Together, these factors can disrupt the fluidity of gameplay, leaving players frustrated and disconnected from the virtual world.

Within the world of online multiplayer gaming, the transmission of data is seldom a seamless journey. Unreliable packets, those mischievous bits of information, can create issues by arriving out of order or disappearing altogether. When packets are out of order, a player might see an opponent magically teleport across the map and back or perform impossible feats. Data loss results in vital game updates never reaching their intended destination, leaving characters and objects frozen in time. Our mission in this chapter is to combat these issues and bring order to the chaos.

In the realm of online multiplayer gaming, a recurring and often frustrating issue that plagues both developers and players alike is the challenge of lag and latency. In this section, we'll talk about these two fundamental aspects of online gaming and shed light on the profound impact they have on a player's experience. As you already discovered in previous chapters, creating a seamless and immersive multiplayer environment requires a nuanced understanding of these concepts.

Now, let's talk about the impact of lag and latency on gameplay. When a player experiences lag, it disrupts the flow of the game and can lead to missed opportunities, frustration, and, in competitive scenarios, unfavorable outcomes. Imagine firing a weapon in an online shooter, only to have the shot register seconds later, long after your target has moved to safety.

Understanding the causes of lag and latency is crucial for effective mitigation. Network congestion, hardware limitations, and geographic distance between players and servers are common causes. **Network congestion** occurs when the data traffic on a network is too high, causing data packets to be delayed or lost. Hardware limitations, such as a slow internet connection or an underpowered computer, can also contribute to latency.

Mitigating lag and latency is a constant challenge for game developers. One strategy is server optimization, where game servers are finely tuned to handle large volumes of data efficiently. Another approach is client-side prediction and interpolation, techniques that help to smooth out gameplay even when there are network delays; we will talk about these in the *Common compensation techniques* section. On top of these, choosing the right network infrastructure, such as **Content Delivery Networks** (**CDNs**), can significantly reduce latency by placing game assets closer to players.

We've peeled back the layers of lag and latency, understanding how these factors impact online multiplayer games. We saw some of their causes and discussed strategies for mitigation, all with the goal of enhancing a player's gaming experience. In the next section, we will talk about issues specific to unreliable packets, which are what we usually use to transfer data over a network in online multiplayer games.

Dealing with unreliable packets

One of the top concerns that developers grapple with when creating online multiplayer games is the reliability of data packets. In this section, we'll see the complexities surrounding unreliable packets, shedding light on the issues they bring to the forefront of online multiplayer games. As you've already gleaned from our discussions, understanding these challenges is core to crafting a smooth and immersive multiplayer gaming experience.

Unreliable packets, as the name suggests, are data packets sent over a network without any guarantee of arrival or order. They're like letters in the wind, reaching their destination only if the conditions are favorable. These packets are used to transmit non-critical data in online games, such as character positions, because they offer lower latency compared to reliable packets, which come with built-in delivery assurances at the expense of potential lag.

One of the primary issues associated with unreliable packets is packet loss. This occurs when packets sent from one player's device fail to reach a server or another player's device. It's like pieces of a puzzle disappearing into thin air, leading to incomplete and inconsistent data. In a fast-paced action game, packet loss can manifest as abrupt character teleportations, vanished projectiles, or inexplicable desynchronization among players.

Another challenge is the out-of-order arrival of packets. In an ideal world, data packets would arrive at their destination in the same order they were sent. However, the unpredictability of network routes can cause packets to arrive out of order, leading to chaos in the game world. Imagine receiving instructions to assemble a piece of furniture, only to receive the steps out of sequence; it's a recipe for confusion and frustration. Usually, in these cases, we only use the latest data and ignore the older ones, as only the most recent information is relevant to the game.

The consequences of unreliable packets can be dire for gameplay. Packet loss and out-of-order arrivals can lead to player disconnection, incorrect character positions, and erratic synchronization among players. For example, a player's character might appear to jump from one location to another due to missing packets. This not only disrupts immersion but also undermines the fairness and integrity of competitive play.

Mitigating the issues brought on by unreliable packets requires a multifaceted approach. Developers often employ techniques such as client-side prediction, where the client makes informed guesses about missing data to maintain a coherent game state. Interpolation, another valuable tool, smooths out the jitters caused by missing packets by smoothly transitioning between known data points.

In this section, we saw that packet loss is a common issue where packets fail to reach their destination, leading to incomplete and inconsistent data. We also saw that some packets may arrive out of order, causing chaos in the game world. These issues can result in player disconnection, incorrect character positions, and erratic synchronization among players. In the next section, we are going to see the most common compensation techniques to solve these and the lag-related issues.

Common compensation techniques

Welcome to the most anticipated section in our journey through the realm of online multiplayer game development. In the previous sections, we unraveled the complexities of networking, synchronization, and the intricacies of dealing with unreliable packets. Now, we stand at a crucial juncture, ready to explore the fascinating world of **interpolation**, **prediction**, and **extrapolation**, a trio of techniques that hold the key to creating seamless and responsive online gaming experiences, or at least to get as close as we can to this Holy Grail.

Picture this – you're in the heat of an intense multiplayer battle, and the stakes couldn't be higher. In the world of online gaming, every second counts, and every move must be precise. But what happens when network latency rears its head, causing a slight delay in transmitting data between players? This is where interpolation, prediction, and extrapolation come to the rescue.

One of the cornerstones of implementing interpolation, prediction, and extrapolation is the integration of physics simulation. In **Godot Engine**, the physics engine plays a crucial role in determining how objects move and interact within the game world. By marrying physics with prediction algorithms, you can create a realistic and responsive gameplay experience that feels in sync with the laws of our virtual universe.

To orchestrate the symphony of data synchronization, we'll take rid of the `Player` node's `MultiplayerSynchronizer` node and employ some *RPC* methods. These functions serve as the conductor of our data orchestra, allowing us to send the necessary information to clients or a server precisely when it's needed. With RPCs, we can trigger the transmission of interpolated, predicted, or extrapolated data, ensuring that all players stay on the same page.

In the next sections, we'll jump into the implementation of interpolation, prediction, and extrapolation in our online multiplayer top-down adventure prototype. By the end, you'll understand how these techniques work together to compensate for network latency. So, fasten your seatbelts, for we are about to navigate the intricacies of smooth and responsive gameplay in the dynamic world of online multiplayer gaming.

Implementing server-side motion

To have a better setup to understand how lag influences the gameplay experience, we are going to make some changes to the `Player` scene and scripts. Instead of allowing the movement to happen on the client side and being synced to the server and other peers, a player will use input events to change the movement of a server's `Spaceship` instance. This will allow us to also decrease the amount of syncing data sent by `MultiplayerSynchronizer`, since now we will have the motion simulated, based on `Spaceship`'s thrusting and rotating states. To do that, let's open the `res://09.prototyping-space-adventure/Actors/Player/Player2D.tscn` scene. Then, follow the following steps:

1. Select the `MultiplayerSynchronizer` node, and in the **Replication** menu, disable `Spaceship`'s position and rotation sync on both the **Spawn** and **Sync** options:

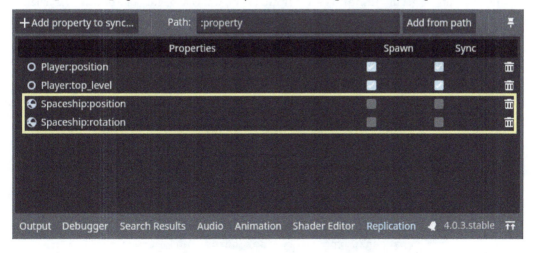

Figure 12.1 – The Player scene's MultiplayerSynchronizer Replication menu,
with the Spaceship position and rotation properties disabled

2. Then, let's open the res://09.prototyping-space- adventure/Actors/Player/ Player2D.gd file, and we will make some changes to the setup_multiplayer() method. The first thing to do here is to remove the line that enables the _physics_process() and _process() callbacks, leaving only _unhandled_input(). We will see why in the following steps:

```
@rpc("any_peer", "call_local")
func setup_multiplayer(player_id):
    var self_id = multiplayer.get_unique_id()
    var is_player = self_id == player_id
    set_process_unhandled_input(is_player)
    camera.enabled = is_player
```

3. Then, we will check whether the current instance isn't the server; if it isn't, we call the make_ current() method, effectively enabling this player's Camera2D node on their game instance:

```
func setup_multiplayer(player_id):
    var self_id = multiplayer.get_unique_id()      var is_player
= self_id == player_id       set_process_unhandled_input(is_
player)      camera.enabled = is_player
    if not multiplayer.is_server():
```

4. Next, we will lay the groundwork the new Spaceship node movement logic, get rid of the _physics_process() callback, and work in the _unhandled_input() callback instead. The whole logic follows the same idea of Weapon2D; Spaceship will have thrusting, direction, and turning variables that we can use to change its movement. Based on the input events we get in _unhandled_input(), we will change the state of these variables. The secret here is that we will use the rpc_id() method to change these states on the server's Spaceship instance.

The following snippet showcases the _unhandled_input() callback after adding this new logic:

```
func _unhandled_input(event):
    if event.is_action_pressed("shoot"):
        weapon.rpc("set_firing", true)
    elif event.is_action_released("shoot"):
        weapon.rpc("set_firing", false)
# Thrusting logic. The spaceship enables its thrust based on if
the `thrust_acti
on` was pressed or released
    if event.is_action_pressed(thrust_action):
        spaceship.rpc_id(1, "set_thrusting", true)
    elif event.is_action_released(thrust_action):
        spaceship.rpc_id(1, "set_thrusting", false)
# Turning logic. If a turning key is just pressed or still
```

```
pressed, the spaceshi
p turns, it only stops turning if neither `turn_left_action` or
`turn_right_action
` are pressed.
    if event.is_action_pressed(turn_left_action):
        spaceship.rpc_id(1, "set_direction", -1)
        spaceship.rpc_id(1, "set_turning", true)
    elif event.is_action_released(turn_left_action):
        if Input.is_action_pressed(turn_right_action):
spaceship.rpc_id(1, "set_direction", 1)
12 Implementing Lag Compensation 8
    else:
        spaceship.rpc_id(1, "set_turning", false)
        spaceship.rpc_id(1, "set_direction", 0)
    if event.is_action_pressed(turn_right_action):
        spaceship.rpc_id(1, "set_direction", 1)
        spaceship.rpc_id(1, "set_turning", true)
    elif event.is_action_released(turn_right_action):
    if Input.is_action_pressed(turn_left_action):
        spaceship.rpc_id(1, "set_direction", -1)
    else:
        spaceship.rpc_id(1, "set_turning", false)
        spaceship.rpc_id(1, "set_direction", 0)
```

5. Now, let's move on to the res://09.prototyping-space- adventure/Objects/
    Spaceship/Spaceship.gd script, where we will implement the variables and methods
    necessary for the aforementioned changes to work. First, let's declare the properties and their
    setter methods:

```
@export var thrusting = false : set = set_thrusting @export var
turning = false : set = set_turning
@export_range(-1, 1, 1) var direction = 0 : set = set_direction
```

6. Then, let's declare these methods; here's the trick – they are *RPCs* that any peer can call, and
    they will be called locally:

```
@rpc("any_peer", "call_local")
func set_thrusting(is_thrusting):
    thrusting = is_thrusting
@rpc("any_peer", "call_local")
func set_turning(is_turning):
    turning = is_turning
@rpc("any_peer", "call_local")
func set_direction(new_direction):
    direction = new_direction
```

7. Then, we will make changes to the `thrust()` and `turn()` methods. The whole idea is that they will receive the delta as an argument now. The `turn()` doesn't need to receive a direction argument anymore, since the direction became a `member` variable:

```
func thrust(delta):
    linear_velocity += (acceleration * delta) * Vector2.RIGHT.
rotated(rotation)
func turn(delta):
    angular_velocity += (direction * turn_torque) * delta
```

8. Finally, we will use the `_physics_process()` callback to call the `thrust()` and `turn()` methods, based on the thrusting and turning variable states:

```
func _physics_process(delta):
if thrusting:
    thrust(delta)
if turning:
    turn(delta)
```

With that, we have everything we need to keep the movement as it was, but now the server is responsible for responding to the player's input instead of being passive to how the `Spaceship` node behaved in the player's game instance. This is important, due to how lag and latency compensation works, as we need an instance of the game to always fall back to if we need to update some data that may have been lost on the network. On top of that, some techniques involve the server side ultimately handling discrepancies. There's an excellent video on *YouTube* called *How to reduce Lag - A Tutorial on Lag Compensation Techniques for Online Games* that explains the role of each side of the connection in lag compensation techniques. This video is available at this link and is highly recommended: https://www.youtube.com/watch?v=2kIgbvl7FRs.

Now that we have this in place, we can start to implement the actual techniques that will help us deal with this issue. In the next section, we will set up our fake lag mechanisms, which are basically `twoTimers`, and see how we can use the `Tween` node to implement *interpolation* in our game so that we can create a fluid motion, based on the sparse `Spaceship` node position and rotation updates.

Bridging the gaps with interpolation

Interpolation is the art of filling in the gaps between received data points. When data packets arrive at irregular intervals due to network latency or packet loss, interpolation ensures that the movement of characters, objects, and projectiles appears smooth and continuous. Imagine it as the magic glue that binds fragmented data, allowing players to witness uninterrupted, fluid motion.

In this section, we will see how we can use the Tween class to interpolate the sparse data we will receive from players. Tween is a specialized class that is used to interpolate values in Godot Engine. We will also use the lerping methods, lerp() and lerp_angle(), to find the correct values to use in the interpolation, especially for Spaceship's rotation angles.

To fake some latency, we will use Timer nodes so that we can see how our interpolation will work in different scenarios. However, ideally, you would use ENetPacketPeer.get_statistic() method passing ENetPacketPeer.PEER_ROUND_TRIP_TIME as argument to get access to the actual network latency. We can access the ENetPacketPeer instance referring to the server's peer connection using multiplayer.multiplayer_peer.get_peer(1) in order to call the get_statistic() method on it. So, to access a player's latency to a server, we can use the following code snippet:

```
# Only clients should get statistics about their connection with the
server, so we don't call that on the server itself.
if not multiplayer.is_server():
    var server_connection = multiplayer.multiplayer_peer.get_peer(1)
    var latency = server_connection.get_statistic(ENetPacketPeer.
PEER_ROUND_TRIP_TIM
E))
    print(latency)
```

That being said, we are going to make some changes to the Player scene and script so that we can implement the interpolation logic and understand how to use this technique. Open the res://09.prototyping-space-adventure/Actors/Player/Player2D.tscn scene and follow the following steps to implement our interpolation logic:

1. Since we are not syncing the Spaceship node's position and rotation properties using the MultiplayerSynchronizer node anymore, we are going to add Timer node to simulate some latency. So, add a new Timer node to the scene, and name it InterpolationTimer.

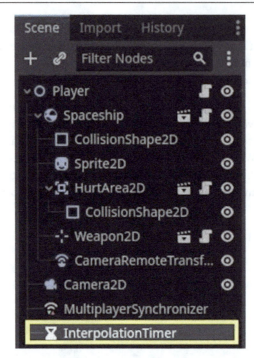

Figure 12.2 – Player's scene node hierarchy with the newly added InterpolationTimer

2. Then, let's set **Process Callback** to **Physics** and **Wait Time** to 0.1. In this context, **Wait Time** represents, in seconds, the latency we want to simulate. A 0.1 wait time would be as high as a 100 ms latency, which is already high enough for players to start noticing some jittering and noticeable delays in their interactions.

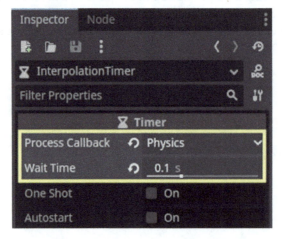

Figure 12.3 – The InterpolationTimer node settings

3. With that, our next step is to connect the `timeout` signal to the `Player` node's script; we
    can create a callback method called `_on_interpolation_timer_timeout()`, like so:

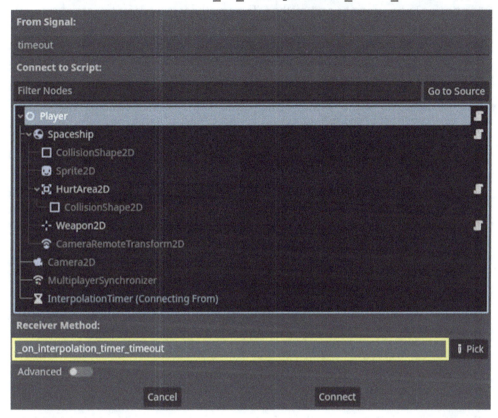

Figure 12.4 – The InterpolationTimer timeout signal connection

4. Then, let's move on to the `res://09.prototyping-space-nadventure/Actors/
    Player/Player2D.gd` script. Here, we will create two new variables to store the previously
    known `Spaceship` node's position and rotation. This will be necessary to interpolate from
    the previous to the newest values moving forward:

    ```
    @onready var previous_position = spaceship.position
    @onready var previous_rotation = spaceship.rotation
    ```

5. Now, in the `_on_interpolation_timer_timeout()` callback, we will make two RPC calls. One to the the `"interpolate_position"` method and the other to the `"interpolate_rotation"` method. These methods will ask for two arguments – the target property (for example, position or rotation), and the duration of the interpolation. In this case, we will use the `InterpolationTimer.wait_time` property as the duration, since this is the time interval between network updates in this context. We will set up these methods in the following steps:

```
func _on_interpolation_timer_timeout():
    rpc("interpolate_position", spaceship.position,
$InterpolationTimer.wait_time)
    rpc("interpolate_rotation", spaceship.rotation,
$InterpolationTimer.wait_time)
```

6. Now, let's declare these methods, starting with `interpolate_position()`. Only the server should be able to call these methods remotely because it's the server that will update these properties, so their `@rpc` annotation should use `"authority"` and `"call_remote"` as options:

```
@rpc("authority", "call_remote")
func interpolate_position(target_position, duration_in_seconds):
```

7. Inside the `interpolate_position()` method, the first thing we will do is create a new Tween instance and store it in a variable, using the `create_tween()` method:

```
@rpc("authority", "call_remote")
func interpolate_position(target_position, duration_in_seconds):
    var tween = create_tween()
```

8. Then, we will use the `lerp()` function to figure out the final value we will use in the interpolation. For the `position` property, this is not as useful, but it will be in the rotation case. However, let's do it this way to maintain some consistency between these functions:

```
@rpc("authority", "call_remote")
func interpolate_position(target_position, duration_in_seconds):
    var tween = create_tween()
```

9. Since we are playing with a body that will run some physics simulation, it's safer to use the `Tween.TWEEN_PROCESS_PHYSICS` mode in `tween` variable so that the interpolation happens during the physics processing. For that, we use the `Tween.set_process_mode()` method:

```
@rpc("authority", "call_remote")
func interpolate_position(target_position, duration_in_seconds):
    var tween = create_tween()
    var final_value = lerp(previous_position, target_position,
1.0)
    tween.set_process_mode(Tween.TWEEN_PROCESS_PHYSICS)
```

10. Then, we can start the actual interpolation; we will store it into a variable called `tweener`, as `Tween.tween_property` returns a `PropertyTween` object that we can use when necessary. In this function, we pass four arguments – the object, the property that's going to be interpolated, the target value, and the duration of the interpolation in seconds:

```
@rpc("authority", "call_remote")
func interpolate_position(target_position, duration_in_seconds):
    var tween = create_tween()
    var final_value = lerp(previous_position, target_position,
1.0)
    tween.set_process_mode(Tween.TWEEN_PROCESS_PHYSICS)
    var tweener = tween.tween_property(spaceship, "position",
final_value, duration_
in_seconds)
```

11. To ensure the interpolation will happen from the previous known value and the most recent one, we will change `tweener`'s starting value, using the `from()` method and passing `previous_position` as an argument:

```
@rpc("authority", "call_remote")
func interpolate_position(target_position, duration_in_seconds):
    var tween = create_tween()
    var final_value = lerp(previous_position, target_position,
1.0)
    tween.set_process_mode(Tween.TWEEN_PROCESS_PHYSICS)
    var tweener = tween.tween_property(spaceship, "position",
final_value, duration_
in_seconds)
    tweener.from(previous_position)
```

12. Then, we update `previous_posistion` to match the now-known most current value, which is our `final_value`:

```
@rpc("authority", "call_remote")
func interpolate_position(target_position, duration_in_
seconds):var tween = create_tween()
    var final_value = lerp(previous_position, target_position,
1.0)
    tween.set_process_mode(Tween.TWEEN_PROCESS_PHYSICS)
    var tweener = tween.tween_property(spaceship, "position",
final_value, duration_
in_seconds)
    tweener.from(previous_position)
    previous_position = final_value
```

13. As for `interpolate_rotation`, we will do the same thing, but this time, we will use the `lerp_angle()` function. This is because interpolating angles is a bit trickier, as we would need to know the shortest path between the starting and target angles. Using this function with a weight of `1.0` provides the final value properly and saves us a lot of time. The whole `interpolate_rotation()` method is very similar to the `interpolate_position()` method but, of course, passing the `previous_rotation` variable instead of the `previous_position` variable. It looks like this:

```
@rpc("authority", "call_remote")
func interpolate_rotation(target_rotation, duration_in_seconds):
    var tween = create_tween()
    var final_value = lerp_angle(previous_rotation, target_
rotation, 1.0)
    tween.set_process_mode(Tween.TWEEN_PROCESS_PHYSICS)
    var tweener = tween.tween_property(spaceship, "rotation",
final_value, duration_
in_seconds)
    tweener.from(previous_rotation)
    previous_rotation = final_value
```

14. Now, we need to start `InterpolationTimer` if the current instance is the connection's server. For that, move to the `setup_multiplayer()` method and add an `else` statement; inside it, start the timer. Don't forget to remove the line that sets up the new instance authority, as from now on, the server itself will always be `Player`'s authority. The `setup_multiplayer()` method should look like this:

```
@rpc("any_peer", "call_local")
func setup_multiplayer(player_id):
    var self_id = multiplayer.get_unique_id() var       is_
player = self_id == player_id        set_process_unhandled_
input(is_player)        camera.enabled = is_player
    if not multiplayer.is_server():
        camera.make_current()
    else:
```

And there we have it – our interpolation logic is ready to smoothly move and rotate our `Spaceship` node, just by getting sparse updates from a server. Note that since we are faking some latency, we are using a fixed interpolation duration. In a more realistic scenario, you'd use the `ENetPacketPeer.PEER_ROUND_TRIP_TIME` statistic as a reference for the actual interpolation duration.

In this section, we saw how we can use the `Timer` node to fake some latency and, by using the `Tween` class, interpolate between two known values for the `Spaceship` node's position and rotation. We also saw how to access some statistics regarding the connection between two peers, especially regarding the latency between clients and a server. However, what happens when we need to keep some consistency in movement while we don't get updates from the server? This is what we will discuss in the next section!

Playing ahead with prediction

Prediction, different from interpolation, is all about playing ahead of the game – quite literally. It involves making informed guesses about an object's future position based on its past behavior. When network delays cause data updates to lag, prediction steps in, ensuring that your character's actions remain responsive and instantaneous, even in the face of network hiccups.

To implement prediction, we are going to use some Newtonian physics to calculate the `Spaceship` node's velocity and project, based on this calculation, where it will likely be in the next tic, and use it to extrapolate its position and rotation moving forward. This will help us prevent the `Spaceship` node from idling.

A core aspect of prediction and extrapolation is that they aim to fix some drawbacks of the interpolation. For instance, from time to time, we need to re-sync the actual `Spaceship` node's position because, otherwise, due to the interpolation duration and potential latency involved, the `Spaceship` node will always be lagged behind, and this can accumulate to a point where the game isn't played in real time anymore. Also, we will use this synchronization time as a reference for the predictions. So, open the `res://09.prototyping-space-adventure/Actors/Player/Player2D.tscn` scene, and let's start implementing the necessary steps:

1. First of all, let's add a new `Timer` node, and name it `SynchronizationTimer`. This one needs to be at a pace greater than `InterpolationTimer`.

Figure 12.5 – Player's scene hierarchy with the SynchronizationTimer node

2.   Then, we will connect `SynchronizationTimer` node's `timeout` signal to the `Player` node's script in a callback, which we can name `_on_synchronization_timer_timeout()`.

Figure 12.6 – The SynchronizationTimer timeout signal connecting to
Player's _on_synchronization_timer_timeout() method

3.   Then, let's open the `res://09.prototyping-space- adventure/Actors/Player/ Player2D.gd` script, and in the `setup_multiplayer()` method, we will also start `SynchronizationTimer` if this instance is the server:

```
@rpc("any_peer", "call_local")
func setup_multiplayer(player_id):
    var self_id = multiplayer.get_unique_id()
    var is_player = self_id == player_id
    set_process_unhandled_input(is_player)
    camera.enabled = is_player
    if not multiplayer.is_server():
        camera.make_current()
    else:
        $InterpolationTimer.start()
        $SynchronizationTimer.start()
```

4. Now, in the `_on_synchronization_timer_timeout()` callback, we will make two *RPCs* – one to a method called `synchronize_position()` and another to a method called `synchronize_rotation()`. We will implement these methods shortly, but for now, just know they ask for a target position and rotation, respectively, and a synchronization tic. For the synchronization tic, we will use the `SynchronizationTimer` node's `wait_time` property as reference:

```
func _on_synchronization_timer_timeout():
    rpc("synchronize_position", spaceship.position,
$SynchronizationTimer.wait_time)
    rpc("synchronize_rotation", spaceship.rotation,
$SynchronizationTimer.wait_time)
```

5. Now, let's start by implementing the `synchronize_position()` method. Only the **Multiplayer Authority** (in this case, the server) should be able to make an RPC to this method, and since the server doesn't need to sync its own `Spaceship` node, it should only call remotely:

```
@rpc("authority", "call_remote")
func synchronize_position(new_position, synchronization_tic):
```

6. Inside this method, we will stop all currently processing Tween instances; note that this approach works in our game because we only have the `interpolate_*()` methods creating Tween instances. If you have other Tween instances running in your game, I recommend storing them in an array and running through them to stop the active ones. We do that to stop the interpolation from continuing as we will set the `Spaceship` node's final position manually:

```
@rpc("authority", "call_remote")
func synchronize_position(new_position, synchronization_tic):
    for tween in get_tree().get_processed_tweens():
        tween.stop()
```

7. Then, we will create a variable to store the future position, based on a prediction we will make, taking the previous and the new positions we just received. We will work on the prediction method later, but for now, just know that it will ask for a new position and how many seconds ahead you want to predict. We will use this prediction to extrapolate movement when we implement extrapolation in the *Gazing into the future with extrapolation* section:

```
@rpc("authority", "call_remote")
func synchronize_position(new_position,        synchronization_
tic):
    for tween in get_tree().get_processed_tweens():
    tween.stop()
    var future_position = predict_position(new_position,
synchronization_tic)
```

8. After that, we can set the `Spaceship` node's position to the new position and update `previous_position` to match the most recent value, so in the next tic, it maintains a reference to the previously updated value:

```
@rpc("authority", "call_remote")
func synchronize_position(new_position, synchronization_tic):
for tween in get_tree().get_processed_tweens():
tween.stop()
var future_position = predict_position(new_position,
synchronization_tic)
spaceship.position = new_position
previous_position = new_position
```

9. As for the `predict_position()` method, it will happen locally on the client's machine, so there's no need to make an RPC here. Let's declare the function's signature and see how we can predict the future with some physics:

```
func predict_position(new_position, seconds_ahead):
```

Inside the `predict_position()` method, we will calculate the distance from the previous position to the new one. We will also calculate the direction from the previous position to the new position so that we have `Vector2` to work with, predicting the movement's velocity:

```
func predict_position(new_position, seconds_ahead):
var distance = previous_position.distance_to(new_position)
var direction = previous_position.direction_to(new_position)
```

10. With that, we will calculate the movement's linear velocity, based on how many seconds ahead we want to predict. We will then set this linear velocity as the `Spaceship.linear_velocity` property so that we it doesn't idle between updates, asthe `Spaceship` node will start moving using this new velocity:

```
func predict_position(new_position, seconds_ahead): var distance
= previous_position.distance_to(new_position) var direction =
previous_position.direction_to(new_position) var linear_velocity
= (direction * distance) / seconds_ahead spaceship.linear_
velocity = linear_velocity
```

11. Finally, we will add the linear velocity to the new position to predict what will be the next position. We will then return this new position so that we can use this value when we decide to extrapolate the `Spaceship` node's movement:

```
func predict_position(new_position, seconds_ahead):
var distance = previous_position.distance_to(new_position)
var direction = previous_position.direction_to(new_position)
var linear_velocity = (direction * distance) / seconds_ahead
spaceship.linear_velocity = linear_velocity
var next_position = new_position + (linear_velocity * seconds_
```

```
ahead)
return next_position
```

12. The logic to predict the rotation will be exactly the same, but take into account that we will use the `learp_angle()` built-in method to figure out the closest angle to extrapolate to. The `synchronize_rotation()` method will look like this:

```
@rpc("authority", "call_remote")
func synchronize_rotation(new_rotation, synchronization_tic):
for tween in get_tree().get_processed_tweens():
tween.stop()
var future_rotation = predict_rotation(new_rotation,
synchronization_tic)
spaceship.rotation = new_rotation
previous_rotation = new_rotation
```

13. The `predict_rotation()` method will look like this:

```
func predict_rotation(new_rotation, seconds_ahead):
var angular_velocity = lerp_angle(previous_rotation, new_
rotation, 1.0) / second
s_ahead
spaceship.angular_velocity = angular_velocity
var next_rotation = spaceship.rotation + (angular_velocity *
seconds_ahead)
return next_rotation
```

With that, we can start to make assumptions of where the `Spaceship` node is likely to be in the near future, based on `SynchronizationTimer` node's tics. However, note that this is a very important available function on the server side, as sometimes, we may want to use it to mitigate lag in *Player* interactions and trigger the right game events. For instance, if we decide to have some **player versus player** (**PvP**) interactions, we may need to predict where a given player's *Spaceship* was when another player fired their gun. This is because, due to latency, the player may have made a guessed shot and landed a hit. However, it is up to the server to decide whether the shot would actually land, given the latency and other aspects.

In this section, we have seen two important techniques to handle lag and latency in online multiplayer games – prediction and synchronization. Prediction involves making informed guesses about an object's future position and rotation, based on its past behavior. To implement prediction, Newtonian physics calculations are used to calculate the `Spaceship` node's velocity and project its likely future position and rotation.

We also saw how to implement the synchronization process, by stopping ongoing `Tween` instances and updating the `Spaceship` node's position and rotation accordingly.

In the next section, we will use the predicted position and rotation to extrapolate `Spaceship` node's movement, both linear and angular, so that if we happen to miss updates, we can at least fake a movement and fix it in the synchronization later if necessary.

Gazing into the future with extrapolation

Extrapolation is the visionary member of the lag compensation trio, gazing into the future to anticipate where objects will be next. By analyzing the current state of a game and the trajectory of objects, extrapolation extends beyond the data you have, offering a glimpse into what lies ahead. This technique is particularly handy for fast-paced games, where a split-second delay can mean the difference between victory and defeat.

The whole idea of extrapolation is that it is an interpolation into the future. Using the predictions we've made, we can create another interpolation, based on some assumptions of where a player is likely to be while we wait for its actual position. This will prevent hiccups and idling between updates. Let's implement our extrapolation algorithm. Open the `res://09.prototyping-space-adventure/Actors/Player/Player2D.gd` script, and follow the following steps:

1. Starting with the function's signature, the `extrapolation_position()` method will ask for the next position and duration in seconds that the extrapolation lasts for. Here, we will use terms similar to the ones in prediction, such as `seconds_ahead`, as we will work with future timing:

    ```
    func extrapolate_position(next_position, seconds_ahead):
    ```

2. This function only happens on the client side, so there's no need to add any RPC annotations to it. Inside this function, we will use a new `Tween` instance to interpolate from the previous known position to the predicted next position, using `seconds_ahead` variable as the duration:

    ```
    func extrapolate_position(next_position, seconds_ahead):
        var tween = create_tween()
        tween.set_process_mode(Tween.TWEEN_PROCESS_PHYSICS)
        var tweener = tween.tween_property(spaceship, "position",
    next_position, seconds_ahead)
        tweener.from(previous_position)
    ```

3. And that's basically it. We will call the `extrapolate_position()` method inside the `synchronize_position()` method right before updating the current and previous positions. Also, we will use the `future_position` variable, which stores the predicted position as an argument for the extrapolated next position:

    ```
    @rpc("authority", "call_remote")
    func synchronize_position(new_position, synchronization_tic):
        for tween in get_tree().get_processed_tweens():
            tween.stop()
    ```

```
        var future_position = predict_position(new_position,
synchronization_tic)
        extrapolate_position(future_position, synchronization_tic)
        spaceship.position = new_position
        previous_position = new_position
```

4. We do the same thing for the `extrapolate_rotation()` method. It should look like this:

```
func extrapolate_rotation(target_rotation, seconds_ahead):
    var tween = create_tween()
    tween.set_process_mode(Tween.TWEEN_PROCESS_PHYSICS)
    var tweener = tween.tween_property(spaceship, "rotation",
target_rotation, secon
ds_ahead)
        tweener.from(previous_rotation)
```

5. The `synchronize_rotation()` method should look like this after adding the line to call the `extrapolate_rotation()` method, using the `future_rotation` variable as an argument:

```
@rpc("authority", "call_remote")
func synchronize_rotation(new_rotation, synchronization_tic):
    for tween in get_tree().get_processed_tweens():
            tween.stop()
        var future_rotation = predict_rotation(new_rotation,
synchronization_tic)
        extrapolate_rotation(future_rotation, synchronization_tic)
        spaceship.rotation = new_rotation
        previous_rotation = new_rotation
```

In this section, you learned about the concept of extrapolation in the context of online multiplayer game development. Extrapolation is a technique that looks into the future to anticipate where objects will be next. By analyzing the current state of a game and the trajectory of objects, extrapolation extends beyond the available data, providing a glimpse into what lies ahead. It is particularly useful in fast-paced games where a split-second delay can significantly impact gameplay. The implementation of extrapolation involves interpolating from the previous known position and rotation to the predicted next position and rotation, using `Tween` instances, with the duration set to the desired time into the future.

Summary

In this chapter, we learned about the issues caused by lag, latency, and packet loss. Then, we saw how to fix them by implementing lag compensation techniques. We explored the concepts of interpolation, prediction, synchronization, and extrapolation to ensure smooth and responsive gameplay, even in the face of network delays.

First, we delved into interpolation, which is the core technique regarding lag compensation. Interpolation helps to fix some drawbacks of latency and sparse data updates by animating between two known values, while actual updates don't arrive. This ensures that the `Spaceship` node won't idling, waiting for new updates from the network. It will smoothly move toward new data, instead of abruptly teleporting to it.

Then, we discussed prediction, which involves making informed guesses about an object's future position, based on its past behavior. By using Newtonian physics calculations, we were able to calculate the spaceship's velocity and project its likely future position and rotation. This helps prevent idle movements and keeps gameplay responsive.

We then explored extrapolation, which extends beyond available data to anticipate where objects will be next. By interpolating from the previous known position and rotation to the predicted next position and rotation, we were able to create smooth movements, even when updates were missed. This technique is particularly useful in fast-paced games where split-second delays can significantly impact gameplay.

By implementing these lag compensation techniques, we can provide players with a seamless and immersive multiplayer gaming experience, even in the presence of network hiccups and delays.

In the next chapter, we will see how we can store some data on a client's machine to reduce the bandwidth used in our game, relying on data that the players already have available on their machines.

13
Caching Data to Decrease Bandwidth

When it comes to reducing bandwidth usage and optimizing network usage in game development, there is a powerful technique that always comes to mind: caching.

Caching solves the question of why we should keep downloading the same data repeatedly when we can download it once, store it somewhere, and reuse it whenever needed. In this chapter, we will delve into caching techniques and learn how to apply them to efficiently download, store, and reuse images and other relevant data. For that, we will use a database that contains image URLs that we are going to download directly from the internet into our players' machines.

To demonstrate the implementation of these caching techniques, we will prototype a new feature in our game project, where players will have the ability to upload custom images for their spaceships. To save time and focus solely on the network aspect of this feature, we will avoid implementing user experience and user interface aspects, leaving those tasks to the talented individuals in our imaginary indie studio. Your role, as a developer, will be to tackle the network-related aspects of this feature and ensure its seamless integration.

In the following screenshot, you can witness the exciting results of this feature in action. Two players are engaging in the game with their own custom spaceship sprites, which have been downloaded from the server. These sprites are sourced from Twemoji, an open source repository of Creative Commons-licensed emojis maintained by Twitter.

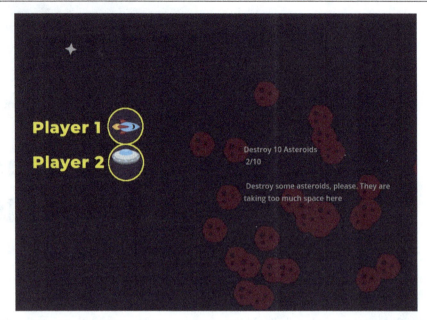

Figure 13.1 – The custom spaceship sprites downloaded from the server for the two players

The topics covered in this chapter are as follows:

- Understanding caching
- Setting up the HTTPRequest node
- Implementing texture caching
- Implementing database caching
- Going further with caching

Technical requirements

It is worth mentioning that this chapter builds upon the concepts presented in *Chapter 10, Debugging and Profiling the Network*, as well as the project developed in *Chapter 9, Creating an Online Adventure Prototype*. Therefore, it is crucial to familiarize yourself with the concepts and techniques discussed in those chapters to fully grasp the optimization methods presented here. We will also build upon the final project from *Chapter 12, Implementing Lag Compensation*, for which you can get the files through the following link:

https://github.com/PacktPublishing/The-Essential-Guide-to-Creating-Multiplayer-Games-with-Godot-4.0/tree/12.prediction-and-interpolation

Moreover, throughout this chapter, it is important to have a basic understanding of how to upload and host content on online services. For instance, we will be downloading files directly from the project's GitHub repository and from a service called ImgBB, a free image-hosting platform. Without understanding the mechanisms behind content hosting and retrieval using direct links, you may encounter difficulties in understanding and implementing the processes we are about to explore.

To enrich your learning experience, I highly recommend downloading the latest version of the Twemoji repository, which can be obtained from `https://github.com/twitter/twemoji`. By exploring this repository, you will gain further insights into how images and other media content can be managed and incorporated into your game development projects.

That said, let's understand what caching is and how we are going to use it in our game project.

Understanding caching

In online multiplayer games, every second counts. Players expect seamless, real-time experiences without interruptions. This is where caching becomes a powerful ally in optimizing game performance. So, what exactly is caching, and why is it crucial for online multiplayer games?

Caching is the process of storing frequently accessed data or resources on a local device or intermediate server. These resources can include images, sound files, 3D models, or even small snippets of code. Instead of fetching these resources from a remote server every time they are needed, the game stores them locally. When a request for these resources arises, the game checks whether it already has a local copy. If it does, it uses the local version, significantly reducing loading times and conserving precious network bandwidth.

The principle behind caching is simple yet effective: if you've used something once, it's likely you'll need it again. In the context of our multiplayer game, this means that the images, sounds, and assets that players download to customize their spaceships can be cached locally on their devices. When another player comes into view, the game can retrieve these assets from the local cache instead of re-downloading them. This creates a more streamlined experience and reduces the stress on the network.

Caching offers numerous advantages in online multiplayer games. The most prominent benefit is an enhanced gaming experience. By using cached resources, players can swiftly interact with others, see their personalized spaceship designs, and engage in combat without significant delays. Reduced loading times translate to more fluid and immersive gameplay. Moreover, this process eases the strain on the server, enabling it to handle more players simultaneously.

Caching is not just about speed; it's also about efficiency. Downloading the same resources repeatedly not only wastes bandwidth but also makes the game less environmentally friendly and potentially costly for players with limited data plans. By caching frequently used resources, the game becomes faster, more eco-friendly, and cost-effective.

In the following sections, we will explore how to leverage the powerful `HTTPRequest` node to implement caching for customized spaceship images. We will dive into the processes of downloading these images and ensuring their availability in each player's cache, enhancing the online multiplayer experience. Stay tuned for a step-by-step guide on how to implement caching in your game.

Setting up the HTTPRequest node

As mentioned in the chapter's introduction, we are going to implement a feature that allows players to use custom sprites on their spaceships.

To prototype this feature, we will download images from a third-party image-hosting service that offers free image hosting. We will accomplish this by using a **Hypertext Transfer Protocol** (**HTTP**) request to retrieve the image file from the third-party servers. Let's delve into the workings of HTTP to fully understand how it operates and grasp the implementation process.

Understanding the HTTP protocol

HTTP serves as the foundation for communication on the **World Wide Web**. It is a protocol that defines the interaction and data exchange between clients and servers. Invented by Tim Berners-Lee in the early 1990s, HTTP was initially designed to facilitate the retrieval of hypertext documents, commonly known as web pages. Over time, it has evolved to support various types of content, including images, videos, and files.

When a client, such as our player, wants to retrieve a resource from a server, it initiates an HTTP request. This request consists of a method that specifies the desired action to be performed on the resource, along with a **Uniform Resource Locator** (**URL**) that indicates the location of the resource. The most-used HTTP methods are GET, POST, PUT, and DELETE. In the context of downloading images, the GET method is what we typically employ.

Upon receiving the HTTP request, the server processes it and prepares an HTTP response. This response contains the requested resource, accompanied by metadata such as the HTTP status code, content type, and content length.

Additionally, the server includes headers in the response to provide further information or instructions to the client.

To download images using HTTP, the client sends a GET request to the server, specifying the URL of the image. The server then processes this request and sends back an HTTP response that contains the image data. The client receives this response and interprets it to display the image to the user.

HTTP operates as a stateless protocol, meaning that each request-response cycle is independent and does not retain any information about previous interactions.

However, mechanisms such as cookies and session management can be employed to maintain state and enable more complex interactions.

In summary, HTTP serves as the protocol that facilitates communication between clients and servers on the web. It enables us to download images and other resources by sending HTTP requests to servers and receiving HTTP responses that contain the requested data. Understanding the workings of HTTP is essential for implementing features such as downloading images in our game project.

The reason we are going to use HTTP requests is because the types of files we want to cache are fairly large compared to the kind of data we usually transfer using **Remote Procedure Calls** (**RPCs**). Remember, the ENet library relies heavily on UDP, which isn't the best option when we want reliable big chunks of data exchange; as UDP aims for high-speed data exchange, packets may arrive at their destination unorganized, and they may not even arrive at the destination at all. When dealing with images, we quickly reach kilobytes or even megabytes of data, so compared to our usual byte exchanges through RPCs and MultiPlayerSynchronizers, it's very heavy content. But HTTP is meant precisely for that. In the upcoming section, we are going to set up the HTTPRequest nodes and a minimal database where we will pair the players' usernames and the URLs of their custom spaceship sprites. This database is intended to go on the server side and be cached in the player's machine later, as we will see in the *Implementing database caching* section.

Setting up the scenes and database

Let's begin setting up the scenes and database.

To create our database, let's open the res://09.prototyping-space-adventure/ folder and create a new text file. You can quickly do that by right-clicking the folder in the **FileSystem** dock. The following screenshot displays the menu that pops up. From there, select **New | TextFile…**

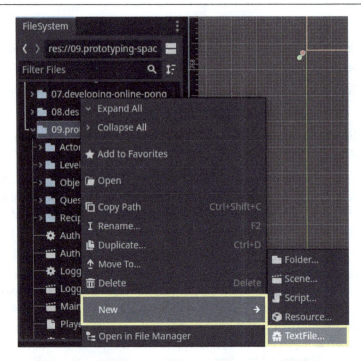

Figure 13.2 – Creating a new text file directly through the FileSystem dock

Then, create a file named `PlayerSpaceships.json`, as shown in the following screenshot:

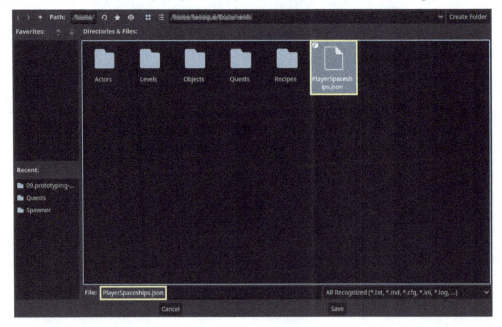

Figure 13.3 – Creating a new text file named PlayerSpaceships.json

Now, regarding the content, we are going to maintain the users from the **FakeDatabase** we created in *Chapter 2, Sending and Receiving Data,* and pair them with the URLs from the images we want to prototype. In this case, I used a service called ImgBB, which allows me to upload the images I rendered from the Twemoji repository. I used the rocket and the saucer emojis as references for these custom sprites. The content of the `PlayerSpaceships.json` file will look like this:

```
{
    "user1": "https://i.ibb.co/KxqzJMp/rocket.png",
    "user2": "https://i.ibb.co/d7BR6hX/saucer.png"
}
```

Note that you can experiment with other images and hosting services. As long as you have the image's direct link, usually pointing to the `.png` file, you will be good.

Now, it's time to set up the `HTTPRequest` nodes. We are going to start by downloading the `PlayerSpaceships.json` file from the server. In our case, this file is hosted on GitHub, but you can store it on another server as long as you have a direct link to the actual database file. In our case you can find it here:

`raw.githubusercontent.com/PacktPublishing/The-Essential-Guide-to-Creating-Multiplayer-Games-with-Godot-4.0/13.caching-data/source/09.prototyping-space-adventure/PlayerSpaceships.json`

With the `PlayerSpaceships.json` file up on the internet, let's see how we can download it to our players' machines with the following steps:

1. Create a new scene and use an `HTTPRequest` node as the root node.

2. Rename it to `SpaceshipsDatabaseDownloadHTTPRequest`, as this node will be responsible for downloading the database from the internet.

3. Attach a new script to this node and save the scene and the script. Here I saved them directly as `res://09.prototyping-space- adventure/SpaceshipsDatabaseDownloadHTTPRequest`.

4. Now, open the script, and let's do the following:

    I.   Create an exported variable for the path to the folder that we will use for caching. Here it's important to use the `user://` data folder path so Godot Engine properly adapts the path depending on the platform the game is running on:

    ```
    extends HTTPRequest

    @export_global_dir var cache_directory = "user://.cache/"
    ```

II. Then, create a new exported variable that should point to where the database file is going to be saved. Let's keep its default filename and put it into the cache folder:

```
extends HTTPRequest

@export_global_dir var cache_directory = "user://.cache/"
@export_global_file var spaceships_database_path =
"user://.  cache/PlayerSpaces
hips.json"
```

III. After that, we can export yet another variable. But now we need to store the link of the location where the PlayerSpaceships.json file will be downloaded from:

```
extends HTTPRequest

@export_global_dir var cache_directory = "user://.cache/"
@export_global_file var spaceships_database_path = "user://.
cache/PlayerSpaces
hips.json"
@export var spaceships_database_link = "https://raw.
githubusercontent.com/PacktPublishing/The-Essential-Guide-
to-Creating-Multiplayer-Games-with-Godot-4.0/13.caching-data/
source/09.prototyping-space-adventure/PlayerSpaceships.json"
```

IV. With that, we can move on to the actual downloading method. Create a new method called download_spaceships_database() and let's start its implementation.

V. The first thing we are going to do here is to check whether there's a cache directory already. If we don't have it yet, we will create it:

```
func download_spaceships_database():
    var directory_access = DirAccess.open(cache_directory)
    if not directory_access:
        DirAccess.make_dir_absolute(cache_directory)
```

VI. Then, we will check whether the PlayerSpaceships.json file exists. If it doesn't, we will start the actual download. The first thing to start the download is to set the file path in the download_file member variable:

```
func download_spaceships_database():
    var directory_access = DirAccess.open(cache_directory)
    if not directory_access:
        DirAccess.make_dir_absolute(cache_directory)
    var file_access = FileAccess.open(spaceships_database_
path, FileAccess.READ)
    if not file_access:
        download_file = spaceships_database_path
```

VII. With `download_file` set, we can make the request to the file. For that, we will use the `request()` method, which asks for a URL. This method uses the `GET` method by default to make requests, which is what we want in this case. But you can change this in the third argument, right after passing some custom headers if you want. We don't have to pass anything other than the URL in our case:

```
func download_spaceships_database():
    var directory_access = DirAccess.open(cache_directory)
    if not directory_access:
        DirAccess.make_dir_absolute(cache_directory)

    var file_access = FileAccess.open(spaceships_database_
path, FileAccess.READ)
    if not file_access:
        download_file = spaceships_database_path
        request(spaceships_database_link)
```

VIII. After that, we need to wait for the request to finish. Remember, since this is an asynchronous procedure, the game needs to wait for it to complete before moving on to any logic that depends on this file:

```
func download_spaceships_database():
    var directory_access = DirAccess.open(cache_directory)
    if not directory_access:
        DirAccess.make_dir_absolute(cache_directory)
    var file_access = FileAccess.open(spaceships_database_
path, FileAccess.READ)
    if not file_access:
        download_file = spaceships_database_path
        request(spaceships_database_link)
        await request_completed
```

After following the preceding steps, we should have our `SpaceshipsDatabaseDownloadHTTPRequest` working. If you want to, you can test it out. For that, call the `download_spaceship_database()` method in the `_ready()` callback and run the scene. After that, if you open the user data folder, you will see the `.cache/` folder and if you enter this folder you should find the `PlayerSpaceships.json` file there. Note that in this case the `.cache/` is a hidden folder, so make sure you can see hidden folders in your file manager. To quickly open the user data folder, you can go to **Project | Open User Data Folder**, as shown in the following screenshot:

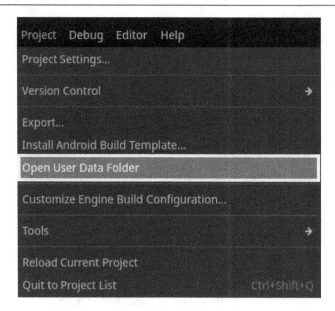

Figure 13.4 - Open User Data Folder from the Editor's Project tab

Now, let's create the `TextureDownloadHTTPRequest` scene so we can use the data in `PlayerSpaceships.json` to effectively download the players' custom spaceship sprites. For that, follow these steps:

1. Create a new scene and use an `HTTPRequest` node as root.

2. Rename it `TextureDownloadHTTPRequest` as this is the one actually responsible for downloading the textures from the internet.

3. Save the scene as `res://09.prototyping-space- adventure/` `TextureDownloadHTTPRequest.tscn` and attach a script to it, then in the script let's do the following:

    I.   Export a variable that should point to the `PlayerSpaceships.json` file in the player's machine, so this path should use the `user://` file path:

    ```
    extends HTTPRequest
    @export_global_file var spaceships_database_file = "user://.
    cache/PlayerSpaces hips.json"
    ```

II.   Create a new method called `download_spaceship()`. This method should receive two arguments, one for the user and another one for the file path where the sprite will be saved:

```
extends HTTPRequest
@export_global_file var spaceships_database_file = "user://.
cache/PlayerSpaces hips.json"

func download_spaceship(user, sprite_file):
```

III. Then, inside this method, we will create a new `Dictionary` called `players_spaceships`, which will start empty but will soon store the content from the `PlayerSpaceships.json` file:

```
func download_spaceship(user, sprite_file):
    var players_spaceships = {}
```

IV.   Now, we are going to check whether the file provided by the `spaceships_database_file` path exists. If it does, we will open it, convert it to a string using `FileAccess.get_as_text()` , and parse it from JSON format to `Dictionary` object format, storing it in the `players_spaceships` variable:

```
func download_spaceship(user, sprite_file):
    var players_spaceships = {}       if FileAccess.file_
exists(spaceships_database_file):
        var file = FileAccess.open(spaceships_database_
file, FileAccess.READ)
        players_spaceships = JSON.parse_string(file.get_as_
text())
```

V.   After that, we can download the `user` sprite based on the URL provided by the `PlayerSpaceships.json` database and store it in the file path provided by the `sprite_file` argument. For that, we will use `HTTPRequest.download_file` and download it using the `HTTPRequest.request()` method. Note that this method returns an error if the request has any issues. Let's store this in an `error` variable to allow other classes to verify whether the request was successful:

```
func download_spaceship(user, sprite_file):
    var players_spaceships = {}
    if FileAccess.file_exists(spaceships_database_file):
        var file = FileAccess.open(spaceships_database_
file, FileAccess.READ)
        players_spaceships = JSON.parse_string(file.get_as_
text())
```

```
        if user in players_spaceships:
            download_file = sprite_file
            var error = request(players_spaceships[user])
```

VI.   Since HTTP requests may take some time to finish downloading the content requested, we need to wait for the `HTTPRequest.request_completed` signal before ending the function and returning the error. Note that if the function doesn't reach any of these conditional statements, we should return `FAILED` to acknowledge to other classes that `spaceships_database_file` doesn't exist or `user` doesn't exist in the database:

```
func download_spaceship(user, sprite_file):
    var players_spaceships = {}
    if FileAccess.file_exists(spaceships_database_file):
        var file = FileAccess.open(spaceships_database_
file, FileAccess.READ)
        players_spaceships = JSON.parse_string(file.get_as_
text())

    if user in players_spaceships:
        download_file = sprite_file
        var error = request(players_spaceships[user])
        await request_completed
        return error
    return FAILED
```

Alright, this ends our journey to set up HTTPRequests. We created the database and the two nodes responsible for working on its data, downloading the database to the players' machines, and downloading the content in the database as well, in this case, the players' spaceships' custom sprites. You can test the scene as well by calling the `download_spaceship()` method using `"user1"` and `"user://.cache/user1_spaceship.png"` as arguments. Just make sure that you've run the `SpaceshipDatabaseDownloadHTTPRequest` scene first so the `PlayersSpaceships.json` file exists in the `user://.cache/` folder. With that, you should see a new image downloaded right in your user data folder! The following screenshot shows what my `user://.cache/` folder looks like:

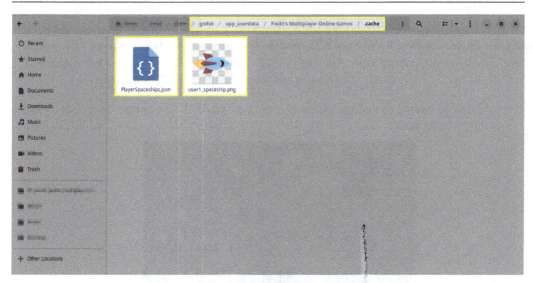

Figure 13.5 – The user data folder with the PlayerSpaceships.json file and user1's custom spaceship sprite

In this section, we learned about the usage of HTTP requests for caching data and downloading images in the user data folder. We saw the difference between UDP and HTTP when it comes to transferring large files such as images and how to use the HTTPRequest node to use the HTTP protocol in Godot and transfer data from the internet to our machine.

In the upcoming section, we will discuss the implementation of texture caching in the game itself, allowing for real-time changes to the players' spaceships. This feature will be achieved by using the HTTPRequest nodes we created and our minimal database to get the URLs of the players' custom spaceship sprites.

Implementing texture caching

In the previous section, we introduced the HTTPRequest node, a built-in solution offered by Godot Engine to make HTTP requests. Then we created TextureDownloadHTTPRequest, which is a custom node specialized in downloading images from our PlayersSpaceship.json database. Now, let's dive deeper into integrating this node into our Player2D class so we can actually use it in our prototype.

In this section, we will create a method to enable the server to change a player's spaceship sprite dynamically. But we won't just load any sprite; we'll fetch the proper file from the user://.cache/ folder we set up in the *Setting up the scenes and database* section. This approach will enhance the customization and interaction in your game, allowing the server to deliver custom sprites to players in real time.

To achieve this, we'll create a method called `load_spaceship()`. This method will play a fundamental role in our implementation. It will be an RPC function that the server can trigger remotely for specific players. Well, let's get started! Open the `res://09.prototyping-space-adventure/Actors/Player/Player2D.tscn` scene and work through the following steps:

1. Let's start by adding an instance of `TextureDownloadHTTPRequest` directly as a child of the `Player` node.

Figure 13.6 – TextureDownloadHTTPRequest instantiated in the Player scene

2. Then, open the `res://09.prototyping-space-adventure/Actors/Player/Player2D.gd` script.

3. Let's store a reference to the `TextureDownloadHTTPRequest` node. We can call it just `http_request` for short:

```
@onready var http_request = $TextureDownloadHTTPRequest
```

4. Then, right below the `setup_multiplayer()` method, let's create the `load_spaceship()` method. It will be an RPC method that should only be remotely callable by the server and should also be called locally, so the server also updates the image. This method receives the user, which loads the spaceship:

```
@rpc("authority", "call_local")
func load_spaceship(user):
```

5. Now, inside this method, we are going to create the file path to the spaceship texture file in the player's machine and store it in a variable called `spaceship_file`. This file path is composed of the `"user://.cache/"` string, followed by the `user` value we got as an argument, and we will append `"_spaceship.png"` to create the proper file path with an image extension:

```
@rpc("authority", "call_local")
func load_spaceship(user):
    var spaceship_file = "user://.cache/" + user + "_spaceship.
png"
```

6. With a proper file path to work with, the next step is to see whether this file already exists because the whole idea of caching data is that, if it already exists in the player's machine, we can load it instead of re-downloading it and reduce resource consumption. The following code checks whether the file exists. If it does, we call the `update_sprite()` method passing `spaceship_file` as an argument. We are going to create the `update_sprite()` method in a moment:

```
@rpc("authority", "call_local")
func load_spaceship(user):
    var spaceship_file = "user://.cache/" + user + "_spaceship.
png"
    if FileAccess.file_exists(spaceship_file):
        update_sprite(spaceship_file)
```

7. Now, `TextureDownloadHTTPRequest` will play its role. If the file doesn't exist in the cache folder, we call the `TextureDownloadHTTPRequest.download_texture()` method and wait for it to finish; remember, it's an asynchronous method. And if this method returns an `OK` error, meaning there's actually no error, we also call the `update_sprite()` method so the game performs the procedures to effectively update the spaceship's sprite. The following code checks whether the `update_sprite()` method returns anything other than `OK`. If the `update_sprite()` method doesn't return `OK`, it means no sprite will be loaded. Since the custom sprite doesn't affect the player's core experience, we can just assume we can keep the default sprite and let the game run:

```
@rpc("authority", "call_local")
func load_spaceship(user):
    var spaceship_file = "user://.cache/" + user + "_spaceship.
png"
```

```
        if FileAccess.file_exists(spaceship_file):
            update_sprite(spaceship_file)
        else:
            if await http_request.download_spaceship(user,
    spaceship_file) == OK:
                update_sprite(spaceship_file)
```

8. Now let's create the `update_sprite()` method. This is the method that effectively changes the `Sprite` node's `Texture` property. It will load the image from the file path using the `Image.load_from_file()` method and turn it into an `ImageTexture` using the `ImageTexture.create_from_image()` method:

```
func update_sprite(spaceship_file):
    var image = Image.load_from_file(spaceship_file)
    var texture = ImageTexture.create_from_image(image)
    $Spaceship/Sprite2D.texture = texture
```

In this section, we saw how to use the `TextureDownloadHTTPRequest` node to download and cache the images from our database. We also saw how to dynamically change a player's spaceship sprite using the cached images, loading them as image resources and turning them into an actual `ImageTexture` that the spaceship's `Sprite2D` node can use.

Now, the missing part of this process is…how we actually get the `PlayersSpaceship.json` database and where we call the `load_spaceship()` method. Well, that's what we are going to do in the next section!

Implementing database caching

With everything in place, it is time to go one step above and work on the `World` scene that is going to glue everything together and run the proper procedures to ensure that all players will see the same custom sprites. We do that in this node because this is the class responsible for setting up everything related to world synchronization, which includes the players' spaceship custom sprites. We will need to make some changes in some core methods to achieve that, but this is for the better.

Since we are working with a prototype, there's no fear of altering function signatures and other core aspects of a class. But note that if this was production-ready, we would call this class "closed" and avoid making core changes like the ones we are about to make. This would keep our game code base consistent and avoid errors. Though the changes we are about to make will mostly extend the class functionalities, we will add an argument to the `World.create_spaceship()` method, breaking the function's contract since it didn't ask for any arguments before. But, as said, this is a prototype, and we have the freedom to tweak things as we want. So, let's open the `res://09.prototyping-space-adventure/Levels/World.tscn` scene and implement the improvements by working through the following steps:

1. First things first, let's add an instance of the `SpaceshipsDatabaseDownloadHTTPRequest` node as a direct child of the `World` node.

Figure 13.7 – SpaceshipsDatabaseDownloadHTTPRequest instantiated in the World scene

2. Then, let's open the `res://09.prototyping-space-adventure/Levels/World.gd` script and start the code by adding a reference to the `SpaceshipsDatabaseDownloadHTTPRequest` node. We can simply call it `http_request` here as well:

```
@onready var http_request =
$SpaceshipsDatabaseDownloadHTTPRequest
```

3. We will also create a new variable to store a **dictionary** that will pair `player_ids` and their `user`. This will allow us to look into this variable to find the proper username related to each player's peer ID so we can easily map them in our database moving forward:

```
var player_users = {}
```

4. Now, in the `_ready()` callback, after waiting briefly for the `0.1` SceneTree's timer to timeout, we will wait for `http_request` to download the spaceships database as well. Remember, this only happens if the instance running the game isn't the server:

```
func _ready():
    if not multiplayer.is_server():
        await(get_tree().create_timer(0.1).timeout)
        await http_request.download_spaceships_database()
```

5. Still in this `if` statement, we will make a small change. When making the RPC to the server's `create_spaceship()` method, we will also pass the player's user. For that, we will use the long-time-forgotten `AuthenticationCredentials` singleton!

```
func _ready():
    if not multiplayer.is_server():
        await(get_tree().create_timer(0.1).timeout)
        await http_request.download_spaceships_database()
        rpc_id(1, "sync_world")
        rpc_id(1, "create_spaceship",
AuthenticationCredentials.user)
```

6. Moving on to the `create_spaceship()` method, we need to, of course, change its function signature to support the `user` argument:

```
@rpc("any_peer", "call_remote")
func create_spaceship(user):
```

7.   Then, before adding the `spaceship` as a child of the `Players` node, we will use its name (which is essentially a string version of the player's peer ID) as a key in the `player_users` dictionary and set this key's value to the `user` argument. With that, we effectively paired the player's ID with their username:

```
@rpc("any_peer", "call_remote")
func create_spaceship(user):
    var player_id = multiplayer.get_remote_sender_id()
    var spaceship = preload("res://09.prototyping-space-
adventure/Actors/Player/Play er2D.tscn").instantiate()
    spaceship.name = str(player_id)
    player_users[spaceship.name] = user
```

8.   After making the RPC to the newly instantiated player spaceship's `setup_multiplayer()` method, we will make the magic happen by also making an RPC to the `load_spaceship()` method, triggering the procedures we made in the *Implementing texture caching* section. The updated version of the `create_spaceship()` method will look like this after these changes:

```
@rpc("any_peer", "call_remote")
func create_spaceship(user):
    var player_id = multiplayer.get_remote_sender_id()
    var spaceship = preload("res://09.prototyping-space-
adventure/Actors/Player/Play er2D.tscn").instantiate()
    spaceship.name = str(player_id)
    player_users[spaceship.name] = user
    $Players.add_child(spaceship)
    await(get_tree().create_timer(0.1).timeout)
    spaceship.rpc("setup_multiplayer", player_id)
    spaceship.rpc("load_spaceship", user)
```

Great! With that, whenever a player jumps into the game, their spaceship will change its sprite to the custom sprite the player uploaded, if any. Now, we need to also make this happen to the spaceships that were already in the game world when the player joined. For that, we need to make some changes in the `_on_players_multiplayer_spawner_spawned()` callback. Still in the `res://09. prototyping- space-adventure/Levels/World.gd` script, let's implement the changes by moving on to the `_on_players_multiplayer_spawner_spawned()` method and go through the following steps:

1.   Let's revamp the whole logic of this callback and make it from scratch. We are going to start by creating a variable to store the player's peer ID, which we can get by turning the recently spawned node's name into an integer. Node names are `StringNames` so they are not pure strings, meaning we need to turn them into default strings before turning them into integers:

```
func _on_players_multiplayer_spawner_spawned(node):
    var player_id = int(str(node.name))
```

2. After that, we will make the global RPC to the `setup_multiplayer()` method, just as we did before, but now we pass `player_id` as the argument:

```
func _on_players_multiplayer_spawner_spawned(node):
    var player_id = int(str(node.name))
    node.rpc("setup_multiplayer", player_id)
```

3. Then comes the interesting part. After making this RPC, if the spawned node, in other words the spawned player, happens to be in another player's game instance, instead of the server's instance, we will also make an RPC directly to the server, asking it to sync this spaceship using a method (which we are going to create soon) called `sync_spaceship()`, which receives a player ID as its argument:

```
func _on_players_multiplayer_spawner_spawned(node):
    var player_id = int(str(node.name))
    node.rpc("setup_multiplayer", player_id)
    if not multiplayer.is_server():
        rpc_id(1, "sync_spaceship", player_id)
```

4. Let's move on to the `sync_spaceship()` method now. Let's create this method with an RPC annotation that will allow any peer to remotely call it and it will also be called locally on the server. Remember, this method receives `player_id` as its argument:

```
@rpc("any_peer", "call_local")
func sync_spaceship(player_id):
```

5. Inside the `sync_spaceship()` method, we will first and foremost store a reference to the ID of whoever made the RPC to this function. This will allow us to make the `rpc_id()` method call directly to the player that just joined the game, instead of making a call to every player every time a player joins the game:

```
@rpc("any_peer", "call_local")
func sync_spaceship(player_id):
    var requester = multiplayer.get_remote_sender_id()
```

6. Then, we will find the correct player in the `Players` node children by using the `get_node()` method and appending the `player_id`. Remember, since the `player_id` is also the node's name, this is how we can easily find it among all the `Players` children:

```
@rpc("any_peer", "call_local")
func sync_spaceship(player_id):
    var requester = multiplayer.get_remote_sender_id()
    var node = get_node("Players/%s" % player_id)
```

7. After that, we will find the player's username. For that, we will use `player_id` as a key in `player_users`, which will return us the proper username:

```
@rpc("any_peer", "call_local")
func sync_spaceship(player_id):
    var requester = multiplayer.get_remote_sender_id()
    var node = get_node("Players/%s" % player_id)
    var user = player_users[node.name]
```

8. Finally, with the user in our hands, we can make an RPC to the `load_spaceship()` method on this requester's instance of the node that represents the spaceship spawned on their instance of the game:

```
@rpc("any_peer", "call_local")
func sync_spaceship(player_id):
    var requester = multiplayer.get_remote_sender_id()
    var node = get_node("Players/%s" % player_id)
    var user = player_users[node.name]
    node.rpc_id(requester, "load_spaceship", user)
```

And we did it! With this in place, the game is now able to load, store, and sync players' custom spaceship sprites. If you run the game now you can see the game caching data to decrease bandwidth and downloading and updating the players' spaceship sprites in real time. The following screenshot displays the before and after of implementing this feature.

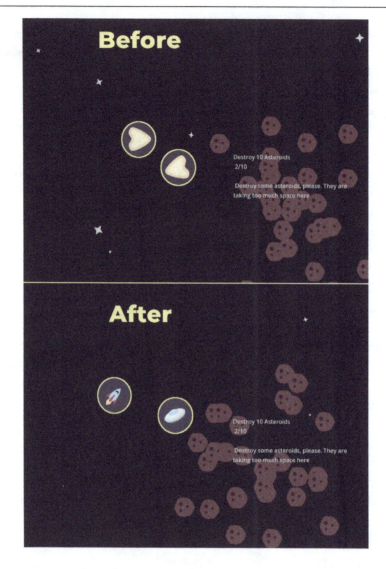

Figure 13.8 – The before and after implementing the custom sprites feature

Something interesting that you can check if you want to measure how much difference caching makes is to remove the statements that check whether the files exist in the `user://.cache/` folder and add a Monitor to monitor the `TextureDownloadHTTPRequest` and `SpaceshipsDatabaseDownloadHTTPRequest` downloaded bytes. For that, you can use the `HTTPRequest.get_downloaded_bytes()` method. For instance, the following code snippet displays how we can create a Monitor in the `TextureDownloadHTTPRequest` node:

```
func _ready():
    var callable = Callable(self, "get_texture_downloaded_bytes")
```

```
    Performance.add_custom_monitor("Network/Texture Download Bytes",
callable)

func get_texture_downloaded_bytes():
    return get_downloaded_bytes()
```

And with that, we can test our game multiple times to simulate players logging in and out of the game and see how the caching impacts each stage. Something you will notice is that after the images have been cached, they are almost instantaneously updated when players log in to the game again. The following screenshot displays the actual impact on the network usage in bytes before and after caching the images in the player's machine:

Figure 13.9 – A comparison between the network consumption the first time the player downloads the textures and after caching the textures in their machine

In this section, we saw how to implement database caching for custom spaceship sprites in our game. The `World` scene is responsible for synchronizing the game world, including the players' spaceship sprites. So, we made some changes to it to achieve the caching we've implemented in previous sections. We saw how to use the `SpaceshipsDatabaseDownloadHTTPRequest` node to download the spaceships database and make RPCs to update players' spaceship sprites and sync spaceship sprites when players join the game. At the end of the section, we used a Monitor and the `HTTPRequest.get_downloaded_bytes()` method to see how caching reduces bandwidth usage and improves network efficiency.

Note the number of bytes saved in *Figure 13.9* with only two textures cached! Imagine, in the long run, how this could impact hundreds and even thousands of players as they log in and out of the game multiple times a day. And we are only talking about textures here. What else can we cache and effectively save hundreds of thousands of network bytes on? Well, that's what we are going to see in the upcoming section.

Going further with caching

Well, as we saw, images are not the only type of data we can download through the internet using the `HTTPRequest` node; for instance, we also downloaded the `PlayersSpaceship.json` file, which is a text file. But we can download pretty much anything using this protocol, provided it is stored in an HTTP page. But sometimes, some files are not stored and made available publicly on an HTTP page that any browser can access. Usually in this type of situation, the backend engineer will create a REST API that we can use to retrieve these files directly from the database where they are stored.

This type of feature demands a physical infrastructure and the development of a custom REST API so we can work with it. Unfortunately, this goes way beyond the scope of this book. But the whole idea is that you can perform an HTTP request using custom headers and a custom URL that resembles an RPC. So, in the very URL itself you would add some parameters that the REST API would interpret as a method call and arguments. It's very similar to what we did in *Chapter 2, Sending and Receiving Data*, and REST APIs usually use JSON-formatted strings as the main data structure as well.

For instance, you can check out a series on my YouTube channel that relies heavily on REST APIs to integrate a third-party service into a game of mine called *Moon Cheeser*. The REST API is provided by *LootLocker* and at some point, I download a whole `PacketScene` from their server to cache it in the player's machine and allow them to always have a copy of their purchased skin. You can check this particular video at the following link:

```
https://youtu.be/w0qz-pJMIBo?si=WF2KH9-FRyO8glVq
```

Well, I'm going to use LootLocker's public API as an example here too. This is the HTTP request I used to retrieve the file associated with the skin the player purchases:

```
curl -X GET "https://api.lootlocker.io/game/v1/assets/
list?count=10&filter=purchasabl e" \-H "x-session-token: your_token_
here"
```

This translates into the following GDScript code, using an HTTPRequest node:

```
var url = "https://api.lootlocker.io/game/v1/assets/
list?count=10&filter=purchasable"
var header = ["Content-Type: application/json", "x-session-token: %s"
% LootLocker.tok
en]
request(url, header)
```

Now comes the interesting part that we haven't explored yet. The HTTPRequest.request_ completed signal is emitted when this request is completed and the server gives a response to the client with some very interesting data, including the response's body, which usually is a JSON file containing the URL to the file we want. So, you can connect this signal to a signal callback and access the response's body to get access to what the server provides you with regarding the request. This can go from the content of a file itself to a JSON file containing much more information, including the URL to download the file you want.

To avoid going off on a tangent here, I highly recommend you watch the video and understand how we download a resource-intensive file. In this case, I downloaded a custom PackedScene instance together with its dependencies, such as an image necessary to make the purchased skin display properly, and cached it into players' devices.

With this in your hands, you can implement all sorts of caching and save tons of resource usage, both for your players and your server, since it won't be constantly delivering the same files to the same clients repeatedly.

Summary

Well, it's time to wrap this chapter up! In this chapter, we discussed what caching is and how it allows us to save bandwidth. We learned how to implement it for custom spaceship sprites in our game. For that, we saw how we can use the HTTPRequest node to download files over the internet by making HTTP requests. We also implemented a custom Monitor to see how much data we saved throughout a play session by caching textures. Finally, we saw how caching can go beyond image and text files and the possibility of using REST APIs to download all sorts of files using HTTP.

With that, my fellow Godot Engine developers, we reach the end of our journey!

We started as people who didn't know how to send a simple message to another computer, and became fully fledged network engineers ready to create the games of our dreams and allow players to have a shared experience, enjoy their time together, and build communities around the world. Congratulations on completing this journey. You are now capable of one of the most amazing things any human being can do: to connect people together toward a shared goal.

This is my farewell. I'm very proud of what we went through, and I hope you use this power for good. You can definitely expect more from me to come. But for now, that's it.

Thank you so much for reading. Keep developing, and until the next time!

Index

`Packtpub.com`

Subscribe to our online digital library for full access to over 7,000 books and videos, as well as industry leading tools to help you plan your personal development and advance your career. For more information, please visit our website.

Why subscribe?

- Spend less time learning and more time coding with practical eBooks and Videos from over 4,000 industry professionals

- Improve your learning with Skill Plans built especially for you

- Get a free eBook or video every month

- Fully searchable for easy access to vital information

- Copy and paste, print, and bookmark content

Did you know that Packt offers eBook versions of every book published, with PDF and ePub files available? You can upgrade to the eBook version at `packtpub.com` and as a print book customer, you are entitled to a discount on the eBook copy. Get in touch with us at `customercare@packtpub.com` for more details.

At `www.packtpub.com`, you can also read a collection of free technical articles, sign up for a range of free newsletters, and receive exclusive discounts and offers on Packt books and eBooks.

Other Books You May Enjoy

If you enjoyed this book, you may be interested in these other books by Packt:

Godot 4 Game Development Projects.

Chris Bradfield

ISBN: 978-1-80461-040-4

- If you're new to Godot, get started with the game engine and editor
- Learn about the new features of Godot 4.0
- Build games in 2D and 3D using design and coding best practices
- Use Godot's node and scene system to design robust, reusable game objects
- Use GDScript, Godot's built-in scripting language, to create complex game systems
- Implement user interfaces to display information

Godot 4 Game Development Cookbook

Jeff Johnson

ISBN: 978-1-83882-607-9

- Speed up 2D game development with new TileSet and TileMap updates
- Improve 2D and 3D rendering with the Vulkan Renderer
- Master the new animation editor in Godot 4 for advanced game development
- Enhance visuals and performance with visual shaders and the updated shader language
- Import Blender blend files into Godot to optimize your workflow
- Explore new physics system additions for improved realism and behavior of game objects

Packt is searching for authors like you

If you're interested in becoming an author for Packt, please visit `authors.packtpub.com` and apply today. We have worked with thousands of developers and tech professionals, just like you, to help them share their insight with the global tech community. You can make a general application, apply for a specific hot topic that we are recruiting an author for, or submit your own idea.

Hi!

I am Henrique Campos, author of *The Essential Guide to Creating Multiplayer Games with Godot 4.0*. I really hope you enjoyed reading this book and found it useful for increasing your productivity and efficiency.

It would really help me (and other potential readers!) if you could leave a review on Amazon sharing your thoughts on this book.

Go to the link below or scan the QR code to leave your review:

`https://packt.link/r/1803232617`

Your review will help us to understand what's worked well in this book, and what could be improved upon for future editions, so it really is appreciated.

Best wishes,

Henrique Campos

Download a free PDF copy of this book

Thanks for purchasing this book!

Do you like to read on the go but are unable to carry your print books everywhere?

Is your eBook purchase not compatible with the device of your choice?

Don't worry, now with every Packt book you get a DRM-free PDF version of that book at no cost.

Read anywhere, any place, on any device. Search, copy, and paste code from your favorite technical books directly into your application.

The perks don't stop there, you can get exclusive access to discounts, newsletters, and great free content in your inbox daily

Follow these simple steps to get the benefits:

1. Scan the QR code or visit the link below

https://packt.link/free-ebook/9781803232614

2. Submit your proof of purchase
3. That's it! We'll send your free PDF and other benefits to your email directly